NED KELLY
Under the Microscope

Special thanks to the invincible Deb Withers, also thanks to Jodie Lee for editorial advice. Also to Peter Carey for his challenging inscription in my copy of The True History of the Kelly Gang – *'To Craig, who will be the author of the 2nd best Kelly book ever written'.*

And thanks to Ned's family members who reminded me that, love him or loathe him, Ned Kelly has living family who still bear the weight not just of his actions, but how we portray him.

This book is dedicated to Anthony Hill – (1945–2013)

The young Ned Kelly. State Library of Victoria.

NED KELLY
Under the Microscope

Solving the forensic mystery of Ned Kelly's remains

EDITED BY CRAIG CORMICK

PUBLISHING

National Library of Australia Cataloguing-in-Publication entry

Ned Kelly under the microscope/editor Craig Cormick.

9781486301768 (paperback)
9781486301775 (epdf)
9781486301782 (epub)

Kelly, Ned, 1855–1880.
Kelly, Ned, 1855–1880 – Death and burial.
Bushrangers – Australia – Biography.
Forensic sciences – Case studies.
Forensic osteology – Case studies.
Human remains (Archaeology) – Victoria.
Excavations (Archaeology) – Victoria.

Cormick, Craig, editor.

364.155092

Published by
CSIRO Publishing
36 Gardiner Road, Clayton VIC 3168
Private Bag 10, Clayton South VIC 3169
Australia

Telephone: [+613] 9545 8555
Local call: 1300 788 000 (Australia only)
Fax: +61 3 9662 7555
Email: csiropublishing@csiro.au
Website: www.publishing.csiro.au

Front cover: (main) Helmet worn by Ned Kelly. State Library of Victoria; (background) kentoh/Shutterstock.com
Title page: Portrait of Ned Kelly. State Library of Victoria.

Set in Adobe Garamond Pro 10.5/13.5
Edited by Adrienne de Kretser, Righting Writing
Cover design by Andrew Weatherill
Typeset by Desktop Concepts Pty Ltd, Melbourne
Index by Bruce Gillespie
Printed by Ingram Lightning Source

The publisher will donate 10% of proceeds to the Donor Tissue Bank of Victoria.

Feb26_RP_ILS

Foreword

Rob Hulls

Looking back over my time in politics, it's strange to realise that few stories knocked others off a front page like the mystery surrounding Ned Kelly's remains. In fact, only the reunion of Phar Lap's skeleton and his equally famous hide came close in terms of the frisson it sent through Melburnian sensibilities.

While this might lead us to conclude that the people of this city have a morbid curiosity about the dead, I believe it goes deeper than this. Certainly it does in the legal community, where the legitimacy of Kelly's trial is a subject of furious debate; his application to the equivalent of the Attorney-General at the time for a grant to pay for a lawyer was a surprising precursor to legal aid.

Equally, however, what happened to Kelly both before and after his execution is as much a story of Melbourne as of Kelly himself. After all, the decade that witnessed the Kelly saga also saw the city's population double in size, the gold rush having laid the foundations. With Victoria taking shape concurrently with Kelly's notoriety, many citizens developed a sense of ownership in his tale – so much so that, when the bodies of Kelly and others were first exhumed, some students from the Working Men's College (now Royal Melbourne Institute of Technology) felt entitled to souvenirs. Bluestones marking the graves also went missing, while others ended up in the beach wall that reaches from Brighton to Beaumaris. Meanwhile, numerous prominent citizens featured in the list of who might be able to assist in the skull's identification.

It's a mistake, then, to characterise the enduring interest in Kelly purely in terms of 'hero versus outlaw'. Rather, it tells us more about ourselves, tapping into strains of what we perceive, rightly or wrongly, as our national identity. It speaks to that disregard for authority which so many Australians, including me, find appealing. It speaks to an urban population's romantic attachment to the bush. It clamours our fascination with a brutal colonial past.

Just as much, however, this is a story of how far we have come – of how the science so essential in identifying Kelly is also fundamental to our understanding of crime, of how its prosecution is more likely to be conducted fairly and its defence properly funded, and of how we are now sufficiently concerned with human dignity to return it to those flung into a mass grave all those years ago.

This is why I became involved in the VIFM appeal for information. Yes, I was as intrigued as the next bloke. 'Is it Ned's Head or Just Another Dull Skull?' was the refrain. I was also proud, however, of a justice system so different from that which Kelly and his contemporaries experienced. While debate continues about the place of the Kelly legend in our collective discourse, the real story should be about the fact that Victorians are served by mechanisms which seek the truth and that, whatever their status in the public imagination, everyone in contemporary Victoria has the right to due process. That is absolutely a story worth telling.

Contents

Timeline

1854 or 1855: Edward (Ned) Kelly is born in Beveridge, north of Melbourne.

1866: Ned Kelly's father, Red Kelly, dies after serving time in prison for cattle theft.

1869: Ned's first run in with police after being in a fight with a Chinese man. Acquitted.

1870: Arrested for assault.

1871: Arrested for riding a stolen horse and for resisting arrest and fighting with police. He is sentenced to three years in prison. Aged about 16.

April 1878: Ned Kelly goes into hiding after Constable Fitzpatrick accuses Ned and his family of trying to kill him. Fitzpatrick was later dismissed from the police force for drunkenness and perjury.

October 1878: Ned Kelly, his brother Dan and friends Steve Hart and Joe Byrne kill three policemen who have been tracking them, at Stringybark Creek near Mansfield.

December 1878: Ned Kelly and the gang hold up a bank in Euroa.

February 1879: Ned Kelly and the gang rob a bank in Jerilderie.

June 1880: The shootout between police and Ned Kelly and the gang at Glenrowan. Ned Kelly is wounded, captured and arrested. The other three gang members are killed.

October 1880: Ned Kelly is put on trial in Melbourne and sentenced to death.

11 November 1880: Ned Kelly is hanged, a death mask is made, and his body buried in the yard of the Old Melbourne Gaol.

1929: The remains of prisoners are dug up from the Old Melbourne Gaol and transferred to unmarked mass graves at Pentridge Prison. It is presumed that many of Ned Kelly's remains are grabbed by souvenir hunters.

1930s: A skull believed to be Ned Kelly's is sent to the Australian Institute of Anatomy in Canberra.

1940s: The skull is removed from display at the Institute of Anatomy.

December 1952: The skull is rediscovered by the Institute in an old safe, after having been missing for some time.

1971: The skull, believed to be Ned Kelly's, and his death mask are given to the National Trust.

1973: The skull goes on display at the Old Melbourne Gaol's museum.

1978: The skull, believed to be Ned Kelly's, is stolen from the Old Melbourne Gaol's museum. Tom Baxter identifies himself as having custody of the skull.

2002: The remains of an executed inmate burial are found at the Old Melbourne Gaol during redevelopment of the site.

2008: Archaeological dig conducted at Glenrowan.

2009: Final burial site at Pentridge Prison located by Heritage Victoria archaeologists. The prison's map indicates that it includes Ned Kelly's remains.

May 2009: 21 coffins containing the remains of executed prisoners exhumed from the site of Pentridge Prison are admitted to the Victorian Institute of Forensic Medicine's mortuary.

November 2009: The skull stolen from the Old Melbourne Gaol Museum is given to the VIFM for identification. Efforts begin to try and find Ned Kelly among the remains of the prisoners exhumed from Pentridge Prison.

September 2011: The Victorian Government announces that VIFM has identified the remains of Ned Kelly but that the skull long thought to be his, is not his.

January 2013: Ned Kelly is buried for the final time, near family members at the cemetery in Greta, northern Victoria.

Preface

Do we really need another Ned Kelly book?

Craig Cormick

On 27 June 1880, Ned Kelly and the three other members of the Kelly gang donned their home-made metal armour and stepped out of Ann Jones' Inn at Glenrowan to face a hail of police bullets – and stepped into Australian folklore.

But it was more than the iconic armour that captured the collective imagination of the public. It was Ned's fighting spirit. It was his willingness to stand up against injustice. It was his determination to stand firm against overwhelming odds and not surrender.

But is it actually our collective need that defines Ned Kelly in this way? I have met Ned Kelly fans as far afield as northern Queensland and Western Australia, who might sport Ned Kelly tattoos or T-shirts and profess their admiration for him, but in fact know very little about his life. That he was the eldest son of an Irish-immigrant mother, once widowed and once deserted, who had a track record of petty crimes and conflicts with the police that escalated out of everyone's control, is rarely known. That he donned armour and stood up to the police force at Glenrowan is what defines him.

However, we easily overlook the darker sides of his reality – that he took part in killing three policemen, or that he planned to derail a police train in what today would be termed an act of terrorism. But we have always forgiven our heroes – and perhaps a nation that is so fond of cutting down its tall poppies needs some heroes that are sealed up in armour – metaphorical or not.

A search of the National Library of Australia's catalogue lists 570 items under the topic of Ned Kelly, 382 of them books. Which begs the question – do we need another Ned Kelly book? My response is, as long as it has something new to say, then 'Yes, we do'.

And this book is unique. It is not another analysis of the story or a probe into Ned Kelly the man, it is a rigorous look at the science behind investigations into Ned Kelly.

The bulk of the book describes the work done by the Victorian Institute of Forensic Medicine in identifying the remains of Ned Kelly, dug up from Pentridge Prison in 2009, and the vexed question of his skull. Several chapters look at the anthropology, odontology and DNA studies used in identification. There are also chapters on metallurgical analysis of the gang's armour to settle the debate on how it was made, archaeological digs at Pentridge Prison and Glenrowan, medical analysis of Ned's wounds and a historical analysis of the records as they relate to many of these scientific investigations.

Science seeks answers to questions through examining evidence, and this book unpicks some of the Kelly myths, based on the scientific evidence (Was Ned Kelly illiterate? Did Dan

Fact or fiction: Having a Ned Kelly tattoo puts you at risk of dying violently

Depending on how you interpret the forensic data, wearing a Ned Kelly tattoo can be very dangerous, and indicates you are much more likely to die an unnatural death. A study conducted by Dr Roger W. Byard, a Professor of Pathology at the University of Adelaide, found that corpses with Ned Kelly tattoos were much more likely to have died by murder and suicide.

The study identified 20 corpses with Ned Kelly-related tattoos in the autopsy files of Forensic Science South Australia. All of the dead were white males, aged between 20 and 67 years of age. Seventeen of the deaths were classed as unnatural, due to suicide in eight cases, accidents in seven cases and homicide in two cases.

In 2010 there were 1117 adult autopsies conducted at Forensic Science South Australia, which included 559 natural deaths, 169 suicides and 14 homicides, which showed that among the population of those with Ned Kelly tattoos there was a 40% higher incidence of suicide and 10% higher incidence of homicide than in the general population.

The study acknowledged the outlaw appeal of Ned Kelly and recognised that there is a link between anti-social tattoos and high-risk behaviour and violence in society. It also stated:

> *The population studied is also a highly selected one and these findings cannot be used to predict associations in the general community … however, in a forensic mortuary it is recognised that certain subsets of tattoos may identify individuals who have been at particular risk of violent and unnatural deaths.*

Craig Cormick

Source: Byard RW (2011) Ned Kelly tattoos: origins and forensic implications. *Journal of Forensic and Legal Medicine* **18**, 276–279.

and Steve escape the fire at Glenrowan?) although it reinforces a few others (such as how severely Ned was wounded yet kept fighting at Glenrowan).

I hope readers of this book will find it as fascinating to read as it was to edit, and appreciate that there is a place where history, folklore and scientific analysis can co-exist. I also want to thank all the authors for their contributions, their patience with my rewrites, and their generosity in donating any royalties to the Donor Tissue Bank of Victoria.

And I should declare my own interests: a book such as this is something I have long dreamt of working on, as Ned Kelly has long been one of my heroes – but so have Australia's scientists!

Introduction

The Hon. John Coldrey QC

The Victorian Institute of Forensic Medicine (VIFM) has been in existence for over 25 years. During that period it has become a world leader in all aspects of forensic medicine and sciences. Its areas of expertise include pathology and related sciences such as toxicology, molecular biology, odontology, anthropology, clinical forensic medicine, the retrieval of tissue for transplantation, and research focusing on accident and injury prevention.

The VIFM motto, 'Truth Conquers All', is never more apparent than when its doctors and scientists give independent expert evidence in the courtrooms of our nation.

On 11 November 2009, a skull was handed to the VIFM by Mr Tom Baxter, a Western Australian farmer. It had been stolen some 31 years earlier from a display cabinet in the Old Melbourne Gaol. The name 'E. KELLY' was written in ink on the side of the skull. Mr Baxter declined to reveal how he had obtained it.

Was this the real Kelly skull? The Victorian government, through Deputy Premier and Attorney-General Rob Hulls, sought the assistance of Professor Stephen Cordner and the VIFM forensic team to answer this question.

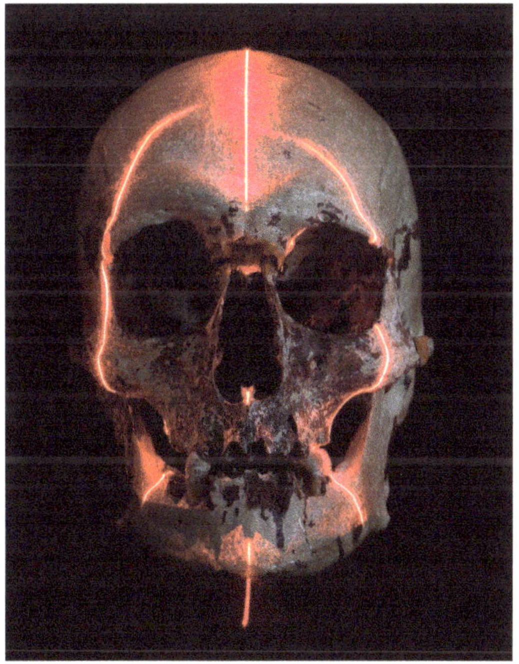

The 'Baxter skull' with CT scan lines. Victorian Institute of Forensic Medicine.

The VIFM conducted craniofacial superimposition, CT scanning and anthropological and DNA tests. Together with historical research, the results demonstrated conclusively that, despite its ink labelling, this was not the skull of the notorious bushranger.

But that was not to be the end of the matter. In 2009 the remains of executed prisoners, exhumed from anonymous graves in Pentridge Prison, had been entrusted by Heritage Victoria to the VIFM. Were the bones of Ned Kelly among them?

A 20-month investigative journey commenced. On 1 September 2011, at an extraordinary media conference, the Victorian Attorney-General, Robert Clark, delivered the verdict. Out of at least 34 sets of remains, many co-mingled, the VIFM team, with great contributions from its collaborators the Argentine Forensic Anthropology Team, had successfully identified Ned Kelly's bones (including a fragment of his skull).

Such is the iconic status of the Ned Kelly narrative that news of the discovery spread world-wide. In addition to coverage throughout Australia and New Zealand, reports were carried as far afield as the BBC World Service, the London *Times* and *Daily Telegraph,* the *Irish Post* and the *New York Times*. Amazingly, the news even appeared on French television and the Al Jazeera network. The then Premier of Victoria, the Honorable Ted Baillieu, informed the Victorian Parliament:

> This was one of the most complex investigations ever undertaken by VIFM. It involved computerised tomography scanning, X-rays, pathology, odontology, anthropology expertise plus extensive historical research and of course DNA analysis. It also involved cooperation with the EAAF laboratory in Argentina, which is a world leader in DNA technology. To be able to identify remains of that age from a gravesite containing [numerous] other bodies is quite remarkable. It demonstrates the world leading expertise Victoria possesses in the field of forensic medicine and forensic science.

But the success of 'The Kelly Project' was also reliant upon the outstanding work of specialists in such disparate disciplines as archaeology, handwriting comparisons, clinical surgery, history (including police and coronial assessments), and public relations management. To give one example, archaeologist Dr. Jeremy Smith of Heritage Victoria was instrumental in the recovery of the 'Baxter skull' and the location of the skeletal remains at Pentridge prison.

In summary, every one of the participants in this complex investigation has made a significant contribution to the creation of a 21st century perspective of the life and death of Ned Kelly.

This is their story.

List of contributors

The Hon. John Coldrey QC

John Coldrey is a retired Justice of the Victorian Supreme Court and chair of the Victorian Institute of Forensic Medicine council. He was Director of Public Prosecutions for Victoria and Director of Legal Services for the Central Land Council in the Northern Territory. In 2004, John was awarded the Gold Medal of the International Society for Reform of Criminal Law, in recognition of his contribution towards criminal law reform.

Adam Ford

Adam Ford is an experienced archaeologist, television presenter and author. He has conducted excavations all over the world and is passionate about investigating and telling the stories of our past. Adam presents the ABC TV series *Who's Been Sleeping In My House?* and has just written his first book.

Carlos Vullo

Carlos María Vullo is a biochemist and PhD in chemical sciences (National University of Córdoba, Argentina). He is the Director of the Forensic DNA Laboratory of the Argentine Forensic Anthropology Team (EAAF), Director of Immunogenetics Laboratory (LIDMO), Córdoba, Argentina. He has published more than 50 scientific papers on immunogenetics and forensic genetics sciences.

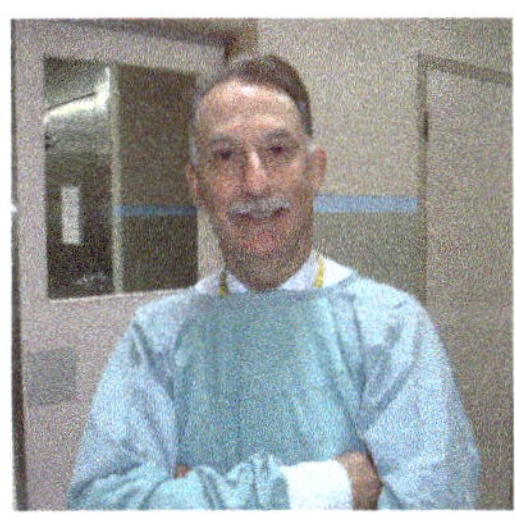

Chris Briggs

Chris Briggs is Consultant Forensic Anthropologist at the Victorian Institute of Forensic Medicine and Associate Professor in the Department of Anatomy and Neuroscience at the University of Melbourne. He has been a member of forensic investigative teams in East Timor, Bali and Christchurch and has published widely in both anatomy and forensic anthropology.

Craig Cormick

Dr Craig Cormick is a science communicator and author. He has published over a dozen books on topics ranging from Antarctica to time travel, and his awards include a Queensland Premier's Literary Award and the ACT Book of the Year Award. He works for CSIRO Education and in 2014 was awarded the Unsung Hero of Science Communications Award by the Australian Science Communicators.

Dadna Hartman

Dr Dadna Hartman completed her PhD in molecular biology at La Trobe University, Victoria, Australia. Since 2008, she has led the Molecular Biology Laboratory at the Victorian Institute of Forensic Medicine, specialising in the DNA analysis of compromised samples, and research activities to improve profiling techniques. She has assisted in the identification of disaster victims, and collaborated with police jurisdictions to solve missing persons and cold case investigations through the provision of DNA analysis.

David Ranson

David Ranson is a medical practitioner and specialist forensic pathologist. He is the Deputy Director of the Victorian Institute of Forensic Medicine and an Associate Professor in the Department of Forensic Medicine at Monash University. He is involved with a wide range of coroners death investigations including homicide case, workplace deaths and deaths in the setting of medical treatment and health care. In addition he is involved with the creative arts side of medicine and forensic pathology, assisting novelists, television shows and film productions.

Dean Wilson

Dean Wilson is Reader in Criminology at Plymouth University in the UK. Prior to that he was a senior lecturer in Criminology at Monash University in Melbourne. He has published a book on the history of policing in Melbourne, as well as many other studies of historical and contemporary criminal justice.

Deb Withers

Deb Withers' career has spanned most areas of the media, as a journalist, publicist and producer. She has written for many publications including the *Age*, the *Herald Sun*, *Woman's Day* and *OK!*, produced hours of television for all networks and represented clients ranging from Red Nose Day to Coles Myer, St Kilda Football Club and the Victorian Institute of Forensic Medicine.

Elizabeth Marsden

Elizabeth Marsden is a historian and museum professional. Previously Collections Manager at the Victoria Police Museum, in 2006 she rediscovered several significant items in the museum's collection related to the Kelly period, including the Thomas McIntyre Collection, Ned Kelly's blood-stained cartridge bag and a small archive of police documents relating to the hunt for the Kelly gang. She is currently co-manager of the Museum Accreditation Program in Victoria.

Fiona Leahy

Fiona Leahy is Senior Medico-legal Adviser at the Victorian Institute of Forensic Medicine, and has a background in legislation and policy development. For the Institute's Kelly Project she complemented her skill base, taking up project management and legal/historical research roles.

Frank McDermott

Frank McDemott was a member of the Department of Surgery at Monash University's Alfred Hospital from 1964 to 1996. In 1973 he joined the Road Trauma Committee of the Royal Australasian College of Surgeons, and was Victorian Chair from 1982 to 1997. His research and promotion led to a zero blood alcohol limit for probationary drivers, and mandatory helmet-wearing by cyclists. With Professor Stephen Cordner he established the Consultative Committee on Road Traffic Fatalities (1992–2005), whose findings led to a new Victorian trauma care system in 2000 and to a marked reduction in preventable trauma deaths.

Gordon Thorogood

Dr Gordon J. Thorogood is a solid-state physicist, specialising in crystallography. He obtained his B. App. Sc. and Masters degrees from the University of Technology, Sydney and a PhD from the School of Chemistry, University of Sydney. He is a council member and the current treasurer of the Australian X-ray Analytical Association. Currently he works at the Australian Nuclear Science and Technology Organisation.

Helen Harris

Helen D. Harris OAM is a professional historian and genealogist, specialising in 19th-century police and criminal records. She holds a Master of Arts in history and is a member of the Professional Historians Association and an Honorary Life Member of the Australian Institute of Genealogical Studies, the Victoria Police Historical Society and the Avoca & District Historical Society.

Helen McKelvie

Helen McKelvie has been Manager of Medico-legal at the Victorian Institute of Forensic Medicine since 1998. Her role involves providing internal legal, policy and compliance advice, quite often in relation to issues around human tissue.

Iain West

Iain West is the Deputy State Coroner, having been appointed to the position in 1993. Prior to this he sat as a magistrate from 1985 to 1993 and before that was a barrister, having been admitted to the Bar in 1975.

Ian Jones

Ian Jones has followed two careers – as a historian and as an award-winning screenwriter, producer and director. He has been a full-time author since 1990. His books include the best-selling biography *Ned Kelly: A Short Life,* as well as *The Fatal Friendship: Ned Kelly, Aaron Sherritt & Joe Byrne.*

Jeremy Smith

Jeremy Smith is Heritage Victoria's Senior Archaeologist and has been a member of the Archaeology Advisory Committee of the Heritage Council since 2002. He has worked on archaeological sites throughout Australia and the Middle East, and on several Ned Kelly related projects at the Old Melbourne Gaol, Pentridge Prison, Stringybark Creek and the Glenrowan siege site.

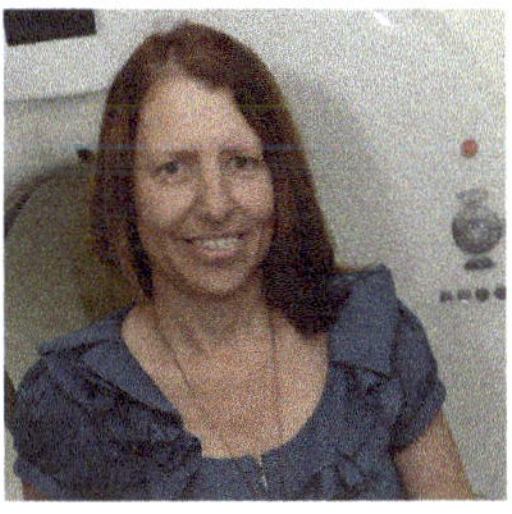

Jodie Leditschke

Jodie Leditschke is Manager of the Forensic Technical Services department at the Victorian Institute of Forensic Medicine. She was chair of the Medical Sciences Advisory Group to the Senior Managers of Australia and New Zealand Forensic Laboratories, is an Adjunct Lecturer at Monash University, a member of the Forensic Analysis Australian Standards Committee and a Founding Fellow of the Faculty of Science of the RCPA.

Lee Franklin

Lee Franklin works as a real estate valuer with Charter Keck Cramer. Lee started his valuation career with the Valuer-General's Office and moved to Charter in 1996. He has had a deep interest in Ned Kelly since childhood.

Malcolm Dodd

Dr Malcolm Dodd is a full-time senior consultant forensic pathologist at the Victorian Institute of Forensic Medicine. He has a special interest in firearms and the interpretation of gunshot injuries and has extensive experience in overseas conflict zones, including East Timor, Kosovo and the Solomon Islands.

Mark Finnane

Mark Finnane is ARC Laureate Fellow at Griffith University, where he works in the ARC Centre of Excellence in Policing and Security. In 2004 his edited volume of the Castieau diaries was published by the National Library of Australia.

Max Esser

Max Esser is an orthopaedic surgeon practising at both Alfred and Cabrini hospitals, Melbourne. He completed his orthopaedic training in the UK and USA and is currently Adjunct Associate Professor in the Department of Surgery at Monash University Alfred Hospital.

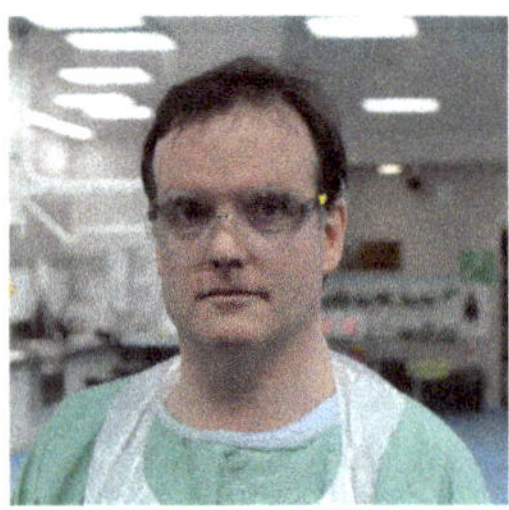

Noel Woodford

Noel Woodford is Head of Forensic Pathology Services at the Victorian Institute of Forensic Medicine, where he has worked for the past 10 years. Prior to his arrival at the Institute he was a consultant Home Office Pathologist and Senior Lecturer in Forensic Pathology in the Department of Forensic Pathology at Sheffield University, UK. He has a Master of Laws degree from the University of Cardiff. His special interests include sudden unexpected natural adult death and radiological imaging as an adjunct to medico-legal death investigation.

Richard Bassed

Dr Richard Bassed is a forensic odontologist and is the consultant in charge of identification services at the Victorian Institute of Forensic Medicine. He has 12 years experience in the forensic field and has a graduate diploma in forensic odontology from Melbourne University and a PhD in forensic medicine from Monash University. He is a Fellow of the Faculty of Oral Pathology within the Royal College of Pathologists of Australasia.

Robert Clark

Robert Clark is the Member for Box Hill and is also the Attorney-General, Minister for Finance and Minister for Industrial Relations in the Victorian government. He was first elected to Parliament in 1988 and has been Member for Box Hill since 1992. From 1992 to 1999 he served as Parliamentary Secretary to the Victorian Treasurer, Alan Stockdale. Before entering parliament, Robert was a solicitor practising in commercial, financial and labour law.

Rob Hulls

Rob Hulls was Attorney-General for Victoria from 1999 to 2010, and during that period held several other ministerial positions. Before entering state parliament, Rob was an Aboriginal legal aid lawyer in Queensland and a member of federal parliament. In 2012 Rob was appointed Director of the Centre for Innovative Justice at RMIT University.

Ronn Taylor

Ronn Taylor has, for almost 25 years, been the forensic sculptor to the Victoria Institute of Forensic Medicine and conducted many workshops nationally and overseas. Ronn's forensic sculpted facial reconstructions assisted in the resolution of many cases of unidentified remains. He has contributed book chapters to several international publications and has appeared in television programs including *Quantum* (ABC) and *Wanted* (Channel 10) as well as featuring in several newspaper and radio interviews regarding forensic sculpture and duplication of skeletal remains.

Soren Blau

Soren Blau (PhD) is the Senior Forensic Anthropologist at the Victorian Institute of Forensic Medicine. She undertakes domestic and international forensic anthropology casework as well as contributing to international training courses on aspects of disaster victim identification and the application of forensic anthropology and archaeology to the investigation of human rights violations. In addition to co-editing a book entitled *Handbook of Forensic Anthropology and Archaeology* (2009), Soren has published numerous peer-reviewed papers on aspects of forensic anthropology.

Stephen Cordner

Stephen Cordner is Professor of Forensic Medicine at Monash University and Director of the Victorian Institute of Forensic Medicine. Stephen has a particular interest in mass casualty management and forensic medical capacity development. Internationally, he has undertaken forensic medical investigations and related missions in the former Yugoslavia, Iraq, Myanmar, the Philippines, East Timor, Fiji and Indonesia. He is Patron of the African Society of Forensic Medicine.

Tahnee Dewhurst

Tahnee Dewhurst is a Senior Forensic Document Examiner with the Victoria Police Forensic Services Department, Melbourne, Australia. She has worked extensively across all facets of questioned document examination, since January 2004, including the provision of expert evidence in courts of law. She is a current PhD candidate with the Department of Human Biosciences, La Trobe University, Melbourne, Australia. She holds the degree of Bachelor of Science (Honours) and a Post Graduate Diploma in Forensic Science (La Trobe University).

Tony Hill

Dr Tony Hill was a forensic odontologist at the Victorian Institute of Forensic Medicine from 1992 until his untimely death in December 2013. Tony helped to identify people who died in mass fatality incidents and was a member of the disaster victim identification team which was involved in the Bali bombing, the Thailand tsunami and the Black Saturday bushfires in Victoria. Tony taught overseas and was the recipient of several awards.

Acknowledgements

Unless stated otherwise, photographs are by, or provided by, the authors.

The photograph of Gordon Thorogood that appears on page xxii is courtesy of Bruce Hudson.

The photograph of Helen Harris that appears on page xxii is courtesy of Suzy Wood.

Thank you to:

- the State Library of Victoria for access to documents and photographs
- the Mitchell Library in Sydney for access to documents
- the unidentified person who made the plaster replica of the skull thought to be Ned Kelly's and the National Museum of Australia for its curation of the replica
- Joanne Griffiths for championing the family's perspective
- all the researchers, workers, writers and many others who contributed in any way to the work that has been documented in this book, and the CSIRO Publishing production team who took the manuscript and turned it into the book it is.

Chapter 1
The arrival of Ned's skull

Helen McKelvie

The skull that was popularly believed to be Ned Kelly's was handed over to the Victorian Institute of Forensic Medicine on 11 November 2009. It was far from the normal type of meeting held at the Institute, and was the trigger for the quest to identify Ned Kelly's skeletal remains.

The curved wall of the Schofield Room (named after Graeme Schofield, founding Professor of Anatomy at Monash University), the conference room at the Victorian Institute of Forensic Medicine, was the perfect vantage point from which to observe the proceedings unfolding around the table. Victoria's leading forensic pathologists from times past, whose portraits had borne witness to countless meetings of more mundane Institute business, were now privileged to view the hand-over of a missing piece of history: a skull inscribed 'E Kelly' that had disappeared from display in 1978 and had been held privately 'in safe-keeping', until now.

The advantage of those witnesses, framed and behind glass, was to avoid the slightly awkward ambience of the occasion. After all, a formal meeting with the holder of skeletal remains, who was admitting them into the coroners' system, was no ordinary occurrence at the Institute. Routine admission procedures were at that time handled by the coroners' registrars and the Institute's mortuary staff. However, this morning's reception had a larger and more diverse cast of characters, including the Institute's Director and key forensic staff.

We were there to meet Mr Tom Baxter, the event's central figure and 'safe-keeper' of the skull since its disappearance. Tom had, through his contact with Dr Jeremy Smith, Senior Archaeologist at Heritage Victoria, chosen the anniversary of Ned Kelly's execution to make his visit to the Institute and bring in the skull (Figs 1.1 and 1.2). This deliberate timing heightened the sense of history-in-the-making, but with no precedent for the event and the assembled experts and other witnesses a little unsure of their roles, the proceedings could be best described as evolving in a somewhat *ad hoc* fashion.

Many of those seated around the table were accustomed to dealing with human remains, of course, but the conference room was not the usual setting for hands-on elements of forensic work. And although dealing with the remains of high-profile individuals was nothing new for the Institute's forensic staff, the prospect of storing and examining the purported skull of our most famous Australian lent the occasion more anticipation than would normally be felt for the receipt of any new 'case'.

Dr Jeremy Smith had accompanied Tom Baxter this morning. We had been getting to know Jeremy during our association with the 'Pentridge remains' (see Chapter 4) and he had told us of his periodic conversations with Tom Baxter. Our anticipation of receiving the skull had therefore been building for several months, following the discovery and

Fig. 1.1: Taking possession of the skull at the Victorian Institute of Forensic Medicine. From left: Dr Jodie Leditshcke, Tom Baxter with the make-up case the skull was presented in, Dr Jeremy Smith. Victorian Institute of Forensic Medicine.

exhumation of executed prisoners' remains from the grounds of the former Pentridge Prison, in February 2008.

The discovery of the first two of three burial pits at the site had seen the commencement of the Institute's collaboration with Heritage Victoria to catalogue the human contents and manage their reinterment. However, with the recovery of further remains from the third pit in February 2009, it was known that Ned Kelly's remains could very possibly be among those now stored at the Institute. But questions regarding the coroner's jurisdiction over the remains and whether exhaustive measures should be taken to try and positively identify Kelly were taking time to resolve.

The expected short-term stay for the Pentridge remains in the VIFM mortuary had blown out, adding to the 'body count' that requires constant storage space choreography by the Institute's mortuary manager, Dr Jodie Leditschke and her staff (see Chapter 10). As the person responsible for the skull's proper handling and safe storage at the Institute, Dr Leditschke was among the reception party for the skull. Also there were Dr Dadna Hartman, manager of the Institute's DNA laboratory, Drs Tony Hill and Richard Bassed, our forensic odontologists whose skills would be called upon during the investigation and who have written chapters in this book.

The skull, not its keeper, was the focus of the meeting. We heard little about Tom Baxter save that he had been a farmer in remote north-western Australia and was now a carpenter in regional Victoria. He had brought a padded make-up case, oval in shape, with a lid, of dated design and a bit worse for wear. It was the receptacle for the skull. The meeting began with introductions and an informal viewing of the make-up case and its contents – which was a suitably aged looking human skull inscribed 'E. Kelly'.

The Institute's Director, Professor Stephen Cordner, led off with some questions for Tom Baxter about the conditions in which the skull had been kept, carefully avoiding direct questioning about the details of how it had come to be in his possession. The potential for reigniting unfinished business associated with the alleged theft of the skull from its National Trust display case (see Chapter 2) was not relevant to the Institute's role, nor something to be pursued by any of those present.

My role in the morning's proceedings was to take notes of Tom Baxter's account and to prepare a statutory declaration for him to sign. This would be used in our attempt to verify the provenance – or authenticity – of the skull and provide a context for the different

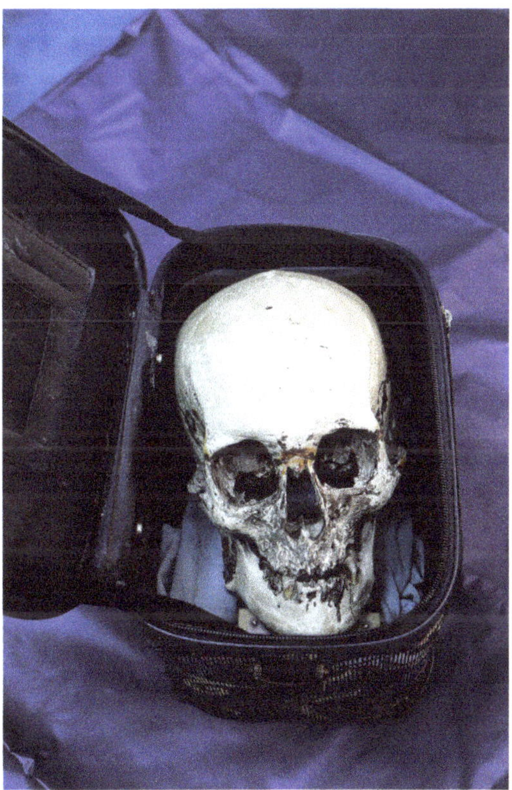

Fig. 1.2: The skull in the beauty case that was given to the Institute by Tom Baxter. Victorian Institute of Forensic Medicine.

scientific techniques that would be applied to it. We were intrigued to hear that its resting place for the past 30 years had been primarily in a hollow log in a paddock on a property in north-west Australia. There it had weathered all sorts of conditions, searing heat and flooding rain, but for the most part the Tupperware container it had been stored in had kept it safe from harm. Tom Baxter did, however, admit to incidents in which the skull had sustained damage, that he carefully repaired with araldite. This included injury sustained during the skull's recent flight from Perth to Melbourne when the make-up case was checked in to the plane's luggage hold rather than carried on board as hand-held baggage; it had perhaps not been handled as carefully as it might have been had the significance of the contents been known by the baggage handlers.

Once we had heard all the details from Tom that might be relevant to a forensic investigation I went upstairs to type up and print out the statutory declaration that Tom would later sign before disappearing again, this time without the skull. I left under the watchful eyes of those forensic experts from long ago, the man who for nearly 30 years had protected what he believed to be Ned Kelly's skull from the indignity of public display, and the more recent forensic scientists who had to keep an open mind both about the provenance of the skull sitting in the make-up case in front of them, and the man who had brought it to them.

Call for public help to solve the mystery of the skull

On Sunday 20 June 2010, the Victorian Attorney-General Rob Hulls and Deputy Director of the VIFM, Associate Professor David Ranson, stood on the gallows where Ned met his death at the Old Melbourne Gaol to ask for the public's assistance in the search for information about the bushranger's execution and burial.

This information would help confirm the authenticity of the skull.

The forensic team was particularly interested in any historical artefacts or records that could possibly confirm the burial place, potential post mortem examination and chain of custody of the skull, which if Ned's was exhumed during the Old Melbourne Gaol excavations in 1929.

'Forensic investigators are half way into the identification of the skull and have discovered areas where they believe the public can be of great assistance,' Mr Hulls said. 'While the forensic science is pivotal in this identification, VIFM needs historical information to determine the provenance of the skull.'

Mr Hulls said VIFM was seeking public help in six areas to locate:

- remains such as bones and teeth reportedly taken as souvenirs by students at the Working Men's College (now RMIT) who were having lunch near the Old Melbourne Gaol when the graves were exhumed in 1929;
- a photo of Alex Talbot, a South Melbourne councillor, holding Ned Kelly's skull, which was mentioned in a newspaper report in 1997. Mr Talbot allegedly took a tooth he believed belonged to Ned Kelly and carried it in his pocket for years. The photo would be great for comparison purposes;
- information on the contractor on the day of Kelly's exhumation, Mr Lee of Lee and Dunn, who took the skull for safe-keeping and then handed it to a state official. Any information from his family could be beneficial;

- information about Maximilian Kreitmayer and his methods of making death masks. Kelly was apparently taken down from the gallows shortly after the execution when Mr Kreitmayer made his death mask. Mr Kreitmayer also had a wax museum in Melbourne;
- information about the work of Sir Colin MacKenzie, the founder of the Australian Institute of Anatomy in Canberra, who was apparently given Kelly's skull to examine after the exhumation in 1929. He had a surgery in St Kilda Rd and was notorious for not keeping records. VIFM would be keen to find anyone who had anything from his work; and
- the missing bluestone block which marked Ned Kelly's grave and has E.K. and a broad arrow on it. Other bluestone blocks that marked the graves at the Old Melbourne Gaol went missing, and some have been found in the beach wall from Brighton to Beaumaris.

'Attempts will be made to obtain a DNA profile although there is only a remote chance that DNA can play a role given the skull is quite degraded and has been handled by an unknown number of people, risking contamination,' Mr Hulls said. 'This is a fantastic historical exercise and I'm sure there is a wealth of interesting information out there, in people's photo albums and sheds, which can help these forensic investigators piece together this tantalising puzzle.'

Although the volume of information forthcoming was small, the quality was excellent.

Chris Ott, the grandson of Alex Talbot, came forward with the tooth his grandfather had souvenired on the day of the excavation in 1929. He also had the photograph of his grandfather holding the skull (from which he took the tooth).

Lee Franklin, grandson of Harry Lee of Lee and Dunn Excavators, also brought forward photographs taken that day at the excavation site in Russell St, and told the team of his grandmother's story of the skull residing on the bedside table the night of the excavation (see p. 17).

With the public's support, the VIFM team was one step closer to solving this mystery.

While at my desk I commented to my legal colleague, Fiona Leahy, on the curious nature of our roles at the Institute. As lawyers employed in a forensic medical institution our working life is unique. It isn't one either of us had consciously aspired to but is one which we nevertheless enjoy and are challenged by every day. And this morning had been even more out of the ordinary.

I didn't know then that this morning's meeting was just the beginning of our Ned Kelly adventure at the Institute or that Tom's return of the skull, so that it might be reunited with any further remains which might be found and together given a dignified reburial, would be the trigger point from which the identification project would really kick off. Within a short time the Attorney-General had offered to provide funding for the project (see p. v) and my colleague Fiona Leahy, who along the way would uncover her love of historical legislation, had drawn up a project plan which would have many iterations. She and the Institute media consultant, Deb Withers, assisted the scientific team in addressing the many and varied issues that arose before the answers became clear, which they describe in other chapters in this book.

Over the many months we worked on the Kelly project the conference room was the venue for numerous meetings, and after that initial uneasy start those framed portraits observed discussions of the ups and downs, frustrations and advances, before the project was finalised. Sadly, I doubt if anyone thought to open the door so they might hear the press conference that took place in the adjacent foyer when the answer to the question about the accuracy of the inscription on the 'Baxter skull' was finally announced to the public.

Chapter 2
The identification of Ned Kelly: a historical perspective

Fiona Leahy and Helen D. Harris

Historical research was undertaken to address the important questions of what happened to Ned Kelly's body after his death, and whether the skull that had been handed into the Victorian Institute of Forensic Medicine could be traced back to the Old Melbourne Gaol. After scouring the archives it was determined that popular history may have been mistaken in believing that his body had been subject to post mortem examination by university anatomists and medical students, but the skull certainly could be traced back to the Gaol. That, however, did not mean it was Ned's skull.

When you visit the Old Melbourne Gaol Museum you can stand at the very spot where Edward 'Ned' Kelly stood, his arms pinioned behind his back, readying himself for death. 'Such is life' or 'So it has come to this' are said to be his last words (see p. 8).

So many things had preceded this moment. In the frenzied two weeks following Kelly's conviction there had been public meetings of his supporters, a petition seeking his reprieve and personal representations to the Governor and the Premier for the commutation of his sentence. Kelly had been visited by family, friends and priests, and had dictated three long letters to the Governor retelling the events leading up to his arrest. But it had all inevitably come to nothing. Just before 10 o'clock on the morning of 11 November 1880 Kelly was met at his cell by the Sheriff, Colonel Rede, and the Governor of the Gaol, Mr J.B. Castieau, and led to the scaffold.

You can imagine the apprehension in the Gaol that morning. The hangman, Elijah Upjohn, was inexperienced, this being his first engagement. And hangings had not always gone smoothly at the Melbourne Gaol. The goal of 'instantaneous death' sometimes eluded prison officials due to human or mechanical failure. They would have been praying for a quick and humane death for the Gaol's most controversial prisoner.

By the eve of his execution, Ned Kelly appeared to have accepted his fate. In his final days he had contemplated what would happen to his body after death. It is at this point in history, the moment of Kelly's death, where our research into the story of Ned Kelly commenced. What had happened to Kelly's body after it was taken down from the scaffold?

We knew that the answer to this question might solve the riddle of the identity of the skull that was presented to the Victorian Institute of Forensic Medicine on 11 November 2009 by Western Australian farmer, Mr Tom Baxter. He believed the skull to be that of Ned Kelly. It had been hidden in a hollow tree stump on his remote property for more than 31 years. Tom Baxter claimed that the skull came into his possession shortly after it was stolen from the Old Melbourne Gaol museum in 1978. Headlines in the *Age* and *Sun News Pictorial* of the time read:

Fact or fiction: Ned Kelly's last words were, 'Such is life!'

What Ned Kelly actually said as his last words is uncertain. Different newspaper reporters stated different things, with papers such as the *Goulburn Herald and Chronicle* reporting that his last words were 'Such is life.' But other papers, including the *Burra Record*, reported that that was his response when the gaol warden advised him of the intended hour of his execution, earlier that morning. The *Argus* reporter who was standing on the gaol floor below the platform, wrote that Ned's last words were, 'Ah, well, I suppose it has come to this.' And the *Sydney Morning Herald* reporter wrote he said, 'Ah, well! It's come to this at last.'

Yet according to the *Illustrated Australian News*, 'He had intended to make a speech, but he uttered no audible sound.'

But one of the closest persons to Ned on the gallows, the gaol warden, wrote in his diary that when Kelly was prompted to say his last words, he opened his mouth and mumbled something that he couldn't hear.

However, just before that when passing through the small garden from his cell to the gallows area, Ned was seen to look at the wheelbarrow that would soon carry his body and at some flowers growing there and say, 'Oh what a pretty garden.' At least according to the reporter from the *Sydney Morning Herald*.

Craig Cormick

> *Ned Kelly skull vanishes*

and

> *Ned loses his head again*

And now we were faced with the possibility that Ned's skull had finally been returned. Or had it? The scientific and medical experts at the Institute are, by training and inclination, a sceptical lot. No-one really believed that this skull could be proven to be that of Ned Kelly, although the puzzle intrigued us all. As it turned out, the more we investigated the skull, the more difficult it was to exclude it from being that of Ned Kelly.

Again the skull prompted a great deal of media interest. An ABC report stated:

> Tom Baxter says he's had that skull for decades, and this week, after ignoring calls for its return for many years, he decided to hand it over to the Victorian Institute of Forensic Medicine. He says his change of heart comes after the discovery at the Old Melbourne Gaol last year of a grave, which is believed to contain Ned Kelly's headless skeleton (ABC AM, 13 November 2009).

Tom Baxter was motivated to hand in the skull as he had been following the story of the skeletal remains that were exhumed from the former Pentridge Prison site, in March 2008 and February 2009. The Pentridge remains were believed to belong to prisoners who had been executed at the Old Melbourne Gaol between 1865 and 1924, and were exhumed from that site when parts of the Gaol were demolished in both 1929 and 1937. The bodies had been reinterred in mass graves at Pentridge Prison. This gravesite later housed bodies of prisoners executed at Pentridge between 1929 and 1967, before capital punishment was formally abolished in Victoria in 1975.

The 45 boxes of human remains were being investigated by the Victorian Institute of Forensic Medicine, as the deaths of these individuals had been reported to the coroner. Among the boxes were thought to be the remains of Ned Kelly. Evidence of this was found on the hand-drawn and undated plan of the unmarked mass graves found in the Pentridge archives, which listed the body of Ned Kelly as being in one of them.

Tom Baxter hoped that the skull could be reunited with the yet to be identified skeletal remains of Ned Kelly.

Ned Kelly – bushranger

The hanging of Ned Kelly caused considerable turmoil in the colony of Victoria in 1880. Kelly and his gang had been on the run from the police for more than two years, during which time they had shot and killed three policemen, as well as a police informer, held up the towns of Euroa and Jerilderie and robbed their banks.

It was evident that Ned Kelly divided opinion during his lifetime as much as he has after his death. Newspapers in every colony followed the Kelly gang's audacious exploits and the relentless police pursuit. Many ordinary citizens volunteered to help the police capture the gang, but a loyal group of friends and sympathisers also frustrated police efforts to track them down.

Fact or fiction: The Kelly gang were tracked by their distinct boot prints

Ned Kelly's boots, from which samples of blood were taken in the hope of finding usable DNA that could be used to match to his bones, were also used to identify Ned and his gang during his outlaw years.

The Kelly gang wore distinctive boots that helped the police in tracking them – with a higher than normal heel, sloping in from the back of the boot. Today it would be described as a Cuban heel but at the time it was called a 'larrikin heel', as it tended to be favoured by flash youths, or larrikins.

At the Royal Commission into the Kelly Outbreak, while being questioned about attempts to discover who had been stealing plough mouldboards, Assistant Commissioner Nicholson stated that Senior Constable Kelly had gone out with two native black trackers and discovered some footmarks.

'One of them described was the footstep of a man with a very small boot, with what is called a "larrikin" heel upon it,' he said.

He also stated, at a separate point in proceedings, 'the blacks explained to him there were traces of a young man with a foot just such as described as Byrne's, with a small foot.'

Other features of larrikin fashion included wearing colourful sashes and fashionable clothes and wearing the chin-straps of their hats under their noses, as is evident in portraits of slain gang informer Aaron Sherritt.

Thomas Carrington, a journalist and artist who interviewed Ned Kelly after he had been captured by the police at Glenrowan, described his dress as: 'He was dressed in the dandy bushman style – yellow cord pants, strapped with slate cross-barred pattern cloth, riding boots, with very thin soles, and very high heels indeed.'

Craig Cormick

The eventual capture and arrest of Ned Kelly, and the death of the three other gang members – Dan Kelly, Steve Hart and Joe Byrne – took place on 28 June 1880 at the hotel of Ann Jones in Glenrowan, a small town in north-east Victoria. The iconic images of Ned Kelly in his armour originate from this, his 'Last Stand' (see Fig. 2.1). Kelly emerged from the chill and foggy bush at dawn, wearing an armour suit fashioned from ploughboards and exchanging gunfire with the gathered police officers. He had been carrying a serious gunshot injury to his left elbow and right foot since the prior evening, and was finally felled by a volley of lead shot fired at his unprotected legs. Following Kelly's arrest the police set fire to the Glenrowan Inn while the bodies of Dan Kelly and Steve Hart were still inside (Sadleir 1913).

After his arrest Kelly was taken to Melbourne and was nursed back to health under the care of Dr Andrew Shields, the Chief Medical Officer and Surgeon of the hospital at the Melbourne Gaol. Upon recovery he was first taken to Beechworth for a committal hearing, then later tried and condemned to death by Justice Redmond Barry for the murder of Constable Lonigan at Stringybark Creek in October 1878.

Fig. 2.1: 'Ned Kelly at Bay'. *Australasian Sketcher*, 3 July 1880. State Library of Victoria.

Fact or fiction: Ned Kelly had a daughter

There is no evidence that Ned had a daughter. Indeed, there is little evidence that he ever had a sweetheart, although several names have been put forward. Historian John McQuilton in his book *The Kelly Outbreak* (1979) cited claims that Ned had married one of Steve Hart's sisters, one of Joe Byrne's sisters and a maid in Deniliquin named Madela!

More likely possibilities of women Ned may have been involved with include Esther 'Ettie' Hart, Steve Hart's sister; Mary Jordan, a barmaid from Jerilderie; Mary Miller, a cousin who helped bring him supplies when his gang was outlawed; or Kate Lloyd, also a cousin and strong supporter of the gang. Adding to the confusion are portrayals of Ned Kelly in novels and films that include a romance angle. In the 1970 movie starring Mick Jagger, before being executed he marries a women in his gaol cell, and in Peter Carey's *The True History of the Kelly Gang* (2000) he has a child to a woman named Mary Hearne, to whom the book is written.

There were clearly women emotionally devoted to Ned. When he was committed in the Beechworth courthouse in August of 1880, newspapers including the *Sydney Morning Herald*, reported:

> *The prisoner also behaved in a singular manner towards a young woman in the gallery, threw kisses to her in a half-reserved sort of way, and she responded at times when she seemed to think no one was noticing. The girl turns out to be an old sweetheart of the outlaw.*

And when he was hanged at the Old Melbourne Gaol in November, several newspapers reported: 'One extraordinary occurrence was that of a woman who, as the clock struck 10, fell on her knees on the footpath, and offered up an audible prayer for the repose of Kelly's soul.'

Craig Cormick

More than 30 000 people signed the petition to the Governor for the reprieve of Ned Kelly (rivalling the 31 000 signatures collected in support of women's suffrage in 1891). But it did not save him. Just 13 days after his conviction, Ned Kelly was judicially hanged at the Melbourne Gaol. There were 21 witnesses to his execution including seven journalists, and a silent crowd of 4000 people gathered outside the Gaol.

In his third and final letter to the Governor, written on the eve of his execution, Ned Kelly appealed to him to 'grant permission for my friends to have my body that they might bury it in consecrated ground' (VPRS 4966, Item 4, Record 13). This wish was not granted and Kelly's body was buried within the precincts of the Melbourne Gaol as required by the *Criminal Law and Practice Statute 1864*.

Such was life

The colony of Victoria was a volatile place following Kelly's execution, particularly in the north-east where his supporters threatened a revolt against the authorities. Superintendent John Sadleir directed that the charred bodies of Kelly's brother Dan and gang member Steve

Hart be given to their families following the siege at Glenrowan, as he feared the consequences if he had not done so. And in the days leading up to Kelly's execution, the Acting Chief Commissioner of Police, Charles Hope Nicolson, directed that police reinforcements be quietly sent to that region where Kelly found most support (VPRS 4965/P0, Unit 2, Item 26).

It was at this tense juncture, just over a week after Kelly's execution, that the *Bendigo Independent* published an article that claimed:

> Ned Kelly's body, (writes our gossipy correspondent) was given over after the execution to the medical men, and a nice mess, I am told, they made of it. The students particularly went in heavily taking part of his body and generally examining every organ. It was a ghastly sight – indeed hardly ever paralleled. I am told that portions of the corpse are now in nearly every 'curiosity' cabinet in Melbourne medical men's places. The skull was taken possession of by one gentleman, and it is probable that he may hereafter enlighten us upon the peculiarities of the great criminal's brain (19 November 1880).

So began the intrigue and mystery that has surrounded Ned Kelly after death: was his body desecrated by the 'medical men'? Did they remove his head and souvenir his skull?

Examination of the skull

As with the Pentridge remains, the skull handed to the Institute by Tom Baxter was classed as 'unidentified human remains found within the state of Victoria', as defined by the *Coroners Act 2008*, and as such the death was reported to the coroner. The former Attorney-General, the Hon. Rob Hulls, recognised the significance of this event and announced that exhaustive tests would be conducted on the surrendered skull to determine if it belonged to Ned Kelly. He announced:

> We now have a unique opportunity, using a combination of historical research and modern technology, to test whether the remains ... are actually those of Ned Kelly (*Herald Sun* 2009).

The Institute was legally obliged to determine for the coroner the 'time since death' relating to the 'Baxter skull' and the Pentridge remains. The coroner has no jurisdiction to investigate deaths that have occurred more than 100 years ago and has discretion to investigate those that have occurred more than 50 years ago.

Having taken custody of the skull, it was carefully examined by our forensic pathologists, anthropologists and odontologists (forensic dentists). Initial scientific examination concluded that it was quite some time since death and that it was likely to belong to a young Caucasian male adult. Significantly, the skull showed no signs of dissection. That meant that it had not been subjected to any post mortem examination involving the brain.

There was no stand-alone scientific means of establishing the time since death relating to the skull short of establishing its identity. The words 'E. Kelly' written on the left side of the skull in black ink may have convinced many who saw the skull that it was Kelly's, but this obviously fell far short of scientific standards of identification (see Fig. 2.2).

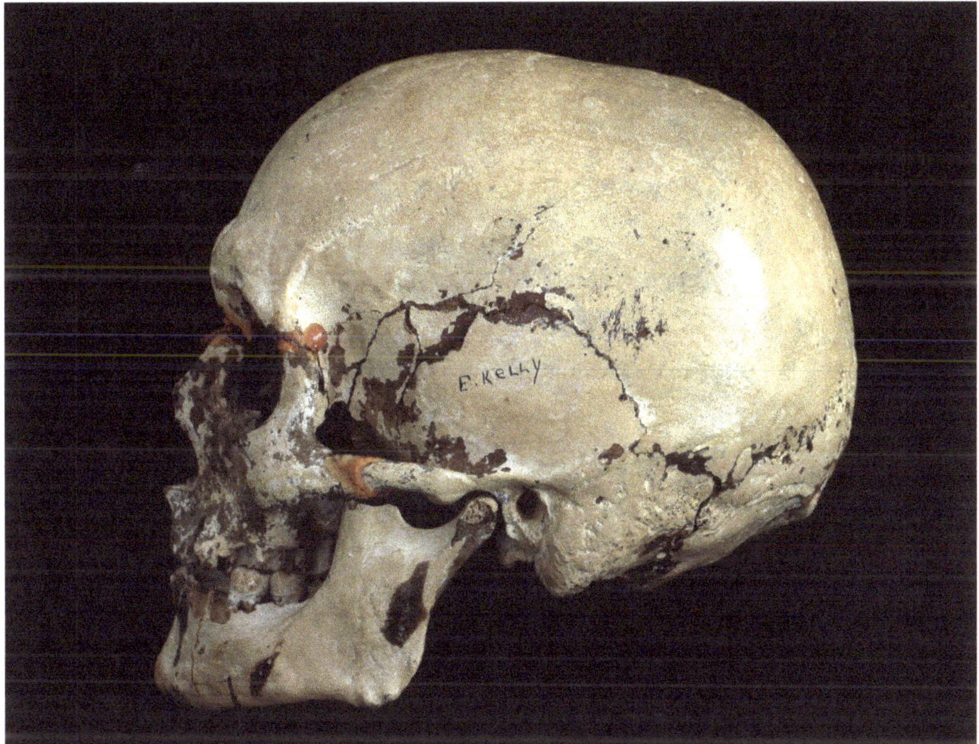

Fig. 2.2: The 'Baxter skull', clearly showing the words 'E. Kelly' hand-written in black ink. Victorian Institute of Forensic Medicine.

The Institute ultimately turned to DNA analysis and tried to obtain suitable mitochondrial DNA to compare it to that of an appropriate family member. But in the beginning we had little confidence of success, given the degraded state of the skull and the risks of contamination from other sources. Added to this, initial attempts at extracting mitochondrial DNA were unsuccessful (see Chapter 9).

Putting together the project team

It is an understatement to say that this investigation was unique in the history of the Institute. We were presented with an unidentified skull believed to belong to Ned Kelly, many boxes of century-old skeletal remains of other executed prisoners, a legal obligation for investigation under the *Coroners Act 2008*, a political interest in the outcome of the investigation and intense media scrutiny given the possibility that we might identify the skull and remains of Ned Kelly.

Our response to this complex task was to establish a project team comprising forensic odontologists, anthropologists, pathologists and molecular biologists. Given the historic nature of this case and its popular appeal, the team also included a lawyer (and historical enthusiast), a consultant historian and media adviser. The project team sought the collaboration of several external experts whose role was pivotal in the final findings.

The two main issues for the historical researchers were:

1. Could we establish the authenticity (or 'provenance', in forensic terms) of the Baxter skull? Could it be linked to the graves of the executed prisoners at the Old Melbourne Gaol, which were exhumed in 1929? Further, could it then be shown to be that of Ned Kelly?
2. Second, was the skull of Ned Kelly subjected to dissection in a post mortem examination of his body, as was suggested in the article published in the *Bendigo Independent* in 1880? If those reported claims were correct, then the Baxter skull could not belong to Ned Kelly as it showed no signs of dissection.

Our first task was to trace the skull from its place in the Institute's mortuary, stepping back through history as far as we could – possibly to Ned Kelly's execution. We began by seeking to determine if it was indeed the skull that had been on display at the Old Melbourne Gaol museum and stolen in 1978.

Archival material provided by the National Trust was invaluable for this. The Trust had been given the skull, together with three others, by the Australian Institute of Anatomy in 1972. The skulls were believed to belong to the executed prisoners Ned Kelly, Frederick Deeming, Martha Needle and Frances Knorr (see p. 113).

A postcard from the Old Melbourne Gaol, with a photo of the skull believed to belong to Ned Kelly on display next to his death mask (see Fig. 2.3), enabled the Institute's forensic odontologists to confirm that the skull on display at the Old Melbourne Gaol was indeed the same as the 'Baxter skull' (see Chapter 6).

The Commonwealth government had established the Australian Institute of Anatomy (AIA) in 1930 to house the collection of fauna and human skeletal remains amassed by its inaugural and sole director, Sir Colin MacKenzie. Sir Colin was a leading orthopaedic surgeon, anatomist and author of books of comparative anatomy of Australian fauna. He is perhaps more recognised for establishing the Healesville Sanctuary for Australian fauna, near Melbourne, originally called the Sir Colin MacKenzie Sanctuary in his honour.

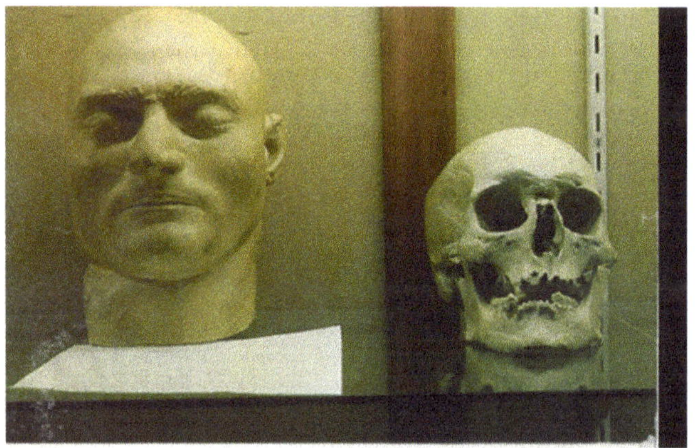

Fig. 2.3: Ned Kelly's death mask and the skull believed to be his, exhibited at the Old Melbourne Gaol Museum before the skull's theft in 1978. National Trust of Australia (Victoria).

Correspondence from the AIA's curator, Mr Robert Stone, provided evidence that the executed prisoners' skulls, and possibly a femur as well, came into the possession of Sir Colin not long after the bodies of executed prisoners were exhumed from the Old Melbourne Gaol in 1929 (National Trust correspondence, January 1969). How and when this transfer of human remains happened is not so clear. While Sir Colin MacKenzie was a man of many talents, record-keeping was not one of them.

We did discover, however, that the AIA created a plaster cast of the skull believed to have belonged to Ned Kelly. This cast is now held by the National Museum of Australia, following the closure of the AIA in 1985. The VIFM's forensic odontologists compared the 'Baxter skull' with the replica and were satisfied that they represented one and the same skull.

Searching for the missing link

Notwithstanding the confirmation that the 'Baxter skull' was the same skull as the replica held by the AIA and regarded as having been that of Ned Kelly, no document trail existed to prove it. There remained a frustrating missing link. It was not possible through an examination of the archival material to complete the chain of custody to prove that the skull held by the AIA was taken from the Old Melbourne Gaol graves of executed prisoners and possibly from Ned Kelly's grave. But then we found a possible lead in a 1997 interview with the former mayor of South Melbourne, Alex Talbot, published in the *Sunday Herald Sun* of 9 March 1997. As a young man Alex Talbot had found work during the Depression with Lee and Dunn Pty Ltd, the construction company engaged by the state government to demolish part of the Melbourne Gaol to make way for an extension to the Working Men's College (now RMIT University).

In the reported interview, he recalled the moment that the graves were discovered in the Old Labour Yard of the Gaol. A crowd had gathered that morning in anticipation that Ned Kelly's coffin might be uncovered. Low along the bluestone wall of the Labour Yard were the carved initials of each of the buried prisoners together with their dates of execution – with the exception of Ned Kelly, whose grave was marked simply with 'EK' and a broad arrow. When the steam shovel scraped open the grave that lay near the EK marker, people rushed forward and souvenired bones from the exposed coffin. Alex Talbot was photographed holding the skull believed to be Kelly's and he took a molar tooth from the skull, which he then kept with the photo (see Figs 2.4, 2.5).

We set out to find Alex Talbot and the tooth from the skull. On 20 June 2010 the Attorney-General made a public appeal for assistance, in particular the whereabouts of Alex Talbot (see p. 4; ABC, 20 June 2010).

Following the issuing of a press release, Chris Ott, grandson of the late Alex Talbot, contacted the Institute (see p. 5). He had the souvenired tooth, preserved in a small wooden box made by his grandfather, and the photo of Alex Talbot holding the skull at the Old Melbourne Gaol in 1929. Chris Ott was keen to discover whether his grandfather's story of Ned Kelly's tooth was actually true, as he had repeated this story many times at primary school when he had brought the Ned Kelly tooth for class news.

We were also contacted by Lee Franklin, whose grandfather was Harry Lee, of Lee and Dunn Pty Ltd, for whom Alex Talbot had worked (see p. 17). Lee Franklin confirmed the

Fig. 2.4: Alex Talbot at the Old Melbourne Gaol in April 1929, holding the skull believed, until recently, to be Kelly's. Chris Ott.

newspaper reports of his grandfather's role in preserving the skull believed to belong to Ned Kelly. Harry Lee had approached the government before the demolition of the Old Labour Yard and voiced his concerns about the possibility of unearthing human remains, only to be incorrectly reassured that there would be nothing found as the bodies had been buried in quicklime and would have disintegrated.

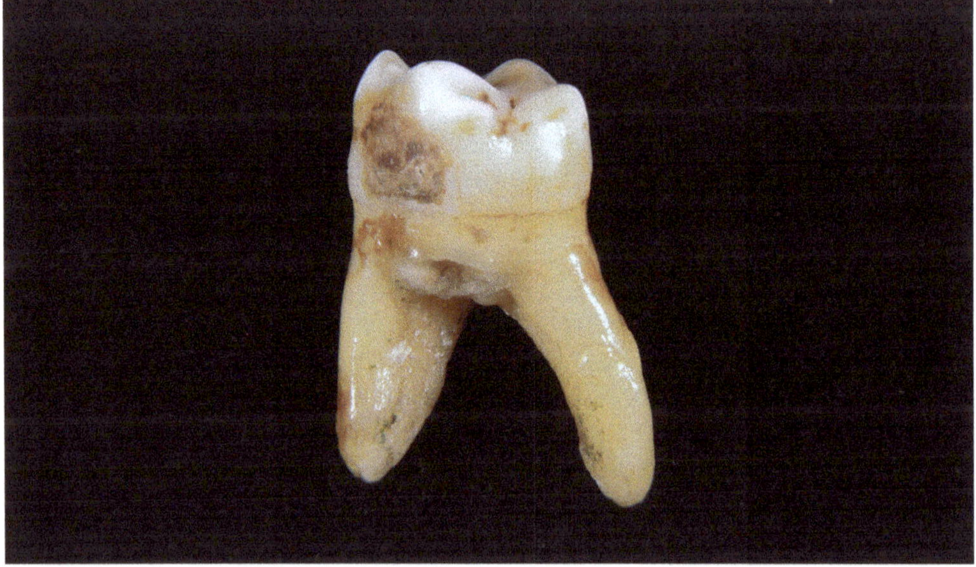

Fig. 2.5: The 'Kelly' tooth, later determined not to be from Ned Kelly. Victorian Institute of Forensic Medicine.

My grandfather dug up Ned Kelly

I am one of those baby boomers who can remember exactly where I was when I heard President Kennedy had been shot (I was in my Nanna's sunroom eating breakfast on a Saturday morning). The sixties were in full swing, the Beatles had just arrived on the scene and I was interested in everything. A book on a bloke named Ned Kelly arrived at Christmas with a picture on the front cover of his remarkable metal helmet. I devoured that book and when I had finished it, I went to tell my Nan all about it.

'You know your grandfather dug him up,' she casually informed me. I didn't know what to make of this. She was a very serious woman and unlikely to be playing a practical joke, but even for a 10-year-old, this seemed a bit far-fetched.

I went home for lunch and said to Mum, 'Nan reckons Grandpa dug up Ned Kelly!' Mum told me that was right and asked what I wanted for lunch! Fortunately, after lunch I went back to see Nan and questioned her at length. She answered my questions, and even showed me a picture of Ned's grave stone, taken at the Old Melbourne Gaol, and related letters from the government and newspaper clippings from the time.

Monday came and I went to school full of this story – but no-one believed me, and of course taking the family heirlooms to school as proof was out of the question.

Nan told me Grandpa had built up a very successful building business, and in 1929 had successfully tendered for the contract to construct police buildings at the Old Melbourne Gaol. Upon inspecting the site he found graves of executed prisoners with their initials carved above each grave in a bluestone wall. Construction could not commence without the remains being exhumed. Grandpa approached the government and was told that the prisoners had all been buried in quicklime and consequently, any remains would be long gone.

So demolition and construction commenced and rough bearings were taken of the site of the one grave everyone was interested in, being that of Ned Kelly. The bluestone wall was then removed and a steam shovel was brought in to dig a long trench to remove the graves. The morning that digging commenced saw a very large crowd of people gathered at the site to watch proceedings. While digging, the workman who had taken the rough bearings said they must be getting close to Ned's grave, and within minutes the steam shovel ripped the

The hand-made box the tooth was stored in. Victorian Institute of Forensic Medicine.

lid off a coffin revealing a skeleton inside. People rushed the grave and in the commotion Grandpa managed to secure the skull, with all other bones being stolen by the onlookers.

We had always believed this to be the skull of Ned Kelly. It was taken to his home in Richmond and kept beside the marital bed for two or three days, before a representative of the government came to pick it up. Nan told me it was taken to the Academy of Science in Canberra where she later viewed it in the mid-sixties.

When Nan died I lost track of the photos and documents for many years, until at a family function a cousin showed me an old leather lunch case. She told me it was Grandpa's and asked if I would like it. I took it home, opened it up and there was the photo of the grave, the newspaper clippings and some correspondence from an embarrassed government that had been seeking to blame Grandpa for the episode. In the end Grandpa employed an undertaker to exhume the remaining graves, one of which was actually Ned's.

Nan told me Grandpa was extremely upset by the whole episode.

Lee Franklin

On 12 April 1929, when the first graves were uncovered and the shambolic scenes of souveniring bones occurred, Harry Lee took the skull believed to belong to Ned Kelly into safe-keeping for a couple of days then handed it to a government official. Lee Franklin has his grandfather's original photographs of the building site at the Old Melbourne Gaol and the EK marker on the bluestone wall. It is very likely that the photo of Alex Talbot holding the skull had been taken by Harry Lee, who must have had a camera on site that day.

Our forensic odontologists examined the molar given to the VIFM by Chris Ott and, to everyone's surprise, the tooth clicked into place in the upper jaw of the skull like Cinderella's foot in her glass slipper – it was a perfect fit. The tooth provided the missing link in the provenance chain, establishing that the skull belonged to one of the prisoners executed at the Melbourne Gaol and buried in the Old Labour Yard.

But was it Ned Kelly?

Despite these historical discrepancies, our research established that the 'Baxter skull' belonged to a prisoner who was buried at the Melbourne Gaol and whose body was exhumed in 1929 and probably reburied at Pentridge Prison. However, we could not exclude the possibility that the skull belonged to Ned Kelly – either by the historical research or the craniofacial photo superimposition (see Chapter 6).

Further, we knew that people were aware of where Ned Kelly had apparently been buried, as his grave was marked with 'EK' (see Fig. 2.6), and were anticipating that this grave would be uncovered on the morning of 12 April 1929, when Lee and Dunn Pty Ltd unearthed the row of coffins.

Had Ned Kelly's body been dissected?

The second important historical question for investigation was whether Ned Kelly had been subjected to a post mortem dissection following his execution. This could have been a formal autopsy for the coroner or an anatomical dissection for teaching purposes.

Fig. 2.6: The initials EK on the wall of the Old Labour Yard at the Old Melbourne Gaol. Victorian Institute of Forensic Medicine. Lee Franklin.

Most Kelly historians, bloggers and enthusiasts did not believe that the skull given to the Institute by Tom Baxter belonged to Ned Kelly, as they relied on the accuracy of the article published in the *Bendigo Independent*, and copied in other country papers, that his body had been subjected to post mortem examination.

We asked ourselves what weight should be placed on this evidence, as it was really just a report from the 'gossipy' correspondent of a regional newspaper that was not reported in the major newspapers of the day. The article's veracity could be tested by the discovery of an autopsy report of the post mortem examination of Kelly's body, which would be expected to be found in the inquest documents recording the coroner's investigation into the death.

Following the hanging of Ned Kelly on the morning of 11 November 1880, Coroner Dr Richard Youl held the inquest into his death and a jury of 12 men inspected his body. The inquest document includes a report from Dr Andrew Shields who was present at the execution. He stated:

> I am the surgeon to this Gaol. I was present in the Gaol this morning when the deceased Edward Kelly was judicially hanged. I have since examined the body of the deceased and found that he was quite dead (VPRS 24, Unit 411, Inquest No. 938 of 1880).

Unfortunately this report gives little indication of whether there was dissection of the body as part of the post mortem examination. Of note, the wording is almost identical to that in the inquest documents for other prisoners executed in the late 19th century. It is telling, however, to compare the inquest documents of prisoners executed at the Gaol with those who died of other causes. For example, in the inquest into the death of James Plunkett who died on 8 January 1881 in the Old Melbourne Gaol, Dr Andrew Shields gave the following evidence to the Coroner:

> I have made a post mortem examination of the body ... the membranes of the brain and the brain itself were highly congested ... other organs except kidneys were healthy (VPRS 24, Unit 417, Inquest No. 16 of 1881).

Dr Shields clearly described a post mortem dissection performed on the body of James Plunkett in his investigation into the cause of death. The different wording of these reports can be explained in part by the legislative authority for the examination of the body following death. Section 16 of the *Medical Practitioners Act 1865* provided the legal authority for a coroner to direct a medical practitioner to examine the body of the deceased. This provision contemplated the dissection of a body as part of the post mortem examination. It provided that:

> the coroner or justice ... may at any time before the termination of the inquest or enquiry direct any legally qualified medical practitioner to perform a post mortem examination of the body of the deceased either with or without an analysis of the stomach or intestines. ...

The judicial hanging of prisoners was, however, regulated by its own statute, *An Act to Regulate the Execution of Criminals 1855* (which was repealed in 1864 and re-enacted as sections 308–314 of the *Criminal Law and Practice Statute 1864*). Section 312 of that Act provided that the:

> coroner of the district in which any gaol may be situated wherein any sentence of death has been carried into execution upon the body of any person shall ... hold an inquest upon the body of such person and the jurors of the jury upon such inquest shall inquire and find whether such sentence was carried into execution.

The Act required that the medical officer of the Gaol be present at each execution and certify in accordance with the wording of the Schedule to the Act that he was in attendance, that he examined the body of the prisoner and that the prisoner was 'hanged by the neck until his body was dead'. There is no specific provision for the coroner to order a post mortem examination that includes dissection. As a matter of public policy, the death of a prisoner from unknown causes would warrant a thorough post mortem investigation in order to exclude the possibility of foul play, whereas the cause of death of a prisoner who has died by judicial hanging was self-evident.

If Ned Kelly had been dissected as part of a post mortem examination we felt that it was unlikely that it was directed by Dr Richard Youl, the coroner. This then raised the question of whether a dissection, if one had been conducted, was lawful. But other legal authority for the dissection of the body of a deceased person did exist at this time, and could be found in the School of Anatomy provisions of the *Medical Practitioners Act 1865*.

In 1864 Professor George Halford commenced teaching anatomy to medical students at the University of Melbourne and in 1876 the Faculty of Medicine was formally established with Professor Halford appointed as inaugural Dean. As the teaching of anatomy requires bodies for dissection, in 1862 the Parliament enacted an *Act for Regulating Schools of Anatomy* which made provision for the legal supply of unclaimed bodies, largely from hospitals.

The number of medical students increased significantly from four students in 1863 to over 100 by 1883 (Victorian Parliamentary Debates 1886), creating a greater demand for the legal supply of bodies to the School of Anatomy. Professor Halford lobbied the Committee of the Benevolent Asylum, asking that unclaimed bodies of those who died in institutions for the working poor be provided for anatomical dissection. His requests were rejected by the Committee, but in 1876 the Parliament passed an amendment to the *Medical Practitioners Act 1865* to enable the bodies of those who died in public institutions – including prisons and lunatic asylums – to be provided to the School of Anatomy.

The legislative amendments to the *Medical Practitioners Act 1865* established an 'opt out' system, which meant the body of any person who died in a prison or lunatic asylum would not undergo anatomical examination in a recognised school of medicine if that person expressed verbally or in writing (before two witnesses) their opposition to such an examination, or if the surviving next of kin required the body to be interred without such examination.

Examining unclaimed bodies

That there were dissections of prisoners in the Old Melbourne Gaol under the 1876 amendments to the *Medical Practioners Act 1865* is evident from the diaries of Mr J.B. Castieau, the Governor of the Gaol between 1869 and 1881. On Thursday 9 August 1877, a year after the amendments to the Act to allow prisoners' bodies to be the subject of anatomical dissection, Castieau noted in his diary that:

> Dr Youl does not take long over an Inquest and it was not many minutes before he had received a verdict and dispersed the Jury. The deceased has been a rather fine looking young woman and when Dr McCrea succeeded in finding out that her body was not claimed by any relative or friend the idea struck him that she would be a 'splendid subject' and made me promise to send her body to the School of Anatomy at the University if no one came to take any interest in it. No friend of the deceased put in an appearance, even the wretch she had been living with made no enquiry, and consequently at about twelve o'clock on Saturday the remains of Ellen Galvin decently coffined were placed in a hearse and conveyed to the School and the care of Dr Halford Professor of Anatomy (Finnane 2004).

Castieau had also described the post mortem examination of an executed prisoner, Edward Feeney, who had been sentenced to death in 1872 for the murder of his friend Charles Marks (see Fig. 2.7). Evidence led in his trial suggested that Feeney and Marks had attempted a double suicide, each holding the muzzle of a pistol to the chest of the other. While Feeney's gun fired, Mark's did not. Castieu wrote:

> There was a Post Mortem held after the usual Inquest and the Medical Students under the direction of Dr Barker revelled in the luxury of a fresh and healthy corpse. 'Oh poor humanity' how small you look upon the dissecting table (Finnane 2004).

Bringing up the bones

The evidence that the skull belonged to a fairly well-defined group of prisoners enabled the Institute's forensic odontologists to compare the skull to the death masks and photos of those prisoners using craniofacial photo superimposition (see Chapter 6). This process provided a means to exclude those prisoners whose faces, as represented by the death masks, did not fit with the skull.

There is, however, a measure of uncertainty about the number of prisoners who were buried in the Old Labour Yard with Ned Kelly, then exhumed in 1929 and reinterred at the Pentridge Prison. According to the records 135 prisoners were judicially hanged at the Old Melbourne Gaol between 1842 and 1924, when closure of the Gaol commenced.

Prior to 1854 the bodies of executed prisoners could be released, upon application, to a family member. People generally believe that this practice stopped following the execution of bushranger George Melville on 3 October 1853. Melville's wife applied for his body and then, to the horror of the authorities, displayed it dressed in flowers and ribbons in the window of her oyster shop in Little Bourke St – charging half a crown for admittance (*South Australian Chronicle and Weekly Mail*, 15 December 1877). From 24 October 1853 to 3 August 1864 the bodies of prisoners executed at the Melbourne Gaol were buried by the government in the Melbourne General Cemetery. In 1865 Parliament amended the law relating to the execution of prisoners and section 315 of *Criminal Law and Practice Statute 1864* required that:

> *the body of every person executed for murder shall be buried within the precincts of the gaol in which he shall have been last confined after conviction and the sentence of the Court shall so direct.*

Between 1865 and 1924 there were 51 prisoners executed at the Old Melbourne Gaol and buried within its grounds. Documentary evidence of these burials can be found in a newly discovered Register of Executed Prisoners presented to the Public Record Office in 2010 (VPRS 16631/P0001, Unit 1). Prisoner John Stacey (executed 1865) through to Angus Murray (executed 1924) are recorded in the register as being 'buried in the Gaol' or 'buried in the Gaol Yard'.

Most of these burials took place in a long narrow yard known as the Old Labour Yard, which was an area in the south-east corner of the Gaol that sloped down to a gateway into Bowen St. In 1894 a visiting journalist described the yard as having:

> *a number of prison tradesmen at work, blacksmiths, carpenters and bootmakers, each in his own little shop, which would be a comfortable little shop if the front of it wasn't really an iron cage. These workers for the Crown look out upon a high wall of rough bluestone and if your attention is drawn to it, but not otherwise, you will notice that close to the foot, and at regular intervals, a space the width of one's hand has been smoothed, and initials cut. They begin near the gate and are creeping up the hill. On the opposite side of the yard the letters 'EK' are alone. Purposely or by chance this distinction is given to the most daring and destructive of Victorian bushrangers. There is no sign of a grave anywhere. In fact it is a road covered with metal sweepings, and every cart that enters the gaol yard passes over scores of buried murderers (*Argus *12 May 1894).*

By 1916 the Old Labour Yard had no room for further burials and the five prisoners who were executed between 1916 and 1924 were buried in the Hospital Yard, directly to the east of the Old Labour Yard. Following the exhumation of the graves in 1929 the Inspector-General of Penal Establishments, Mr Joseph Akeroyd, reported to the Under-Secretary that five bodies 'will remain in the Hospital Yard at present under the control of the Police Department' (VPRS 3992, Unit 1858, File W4499).

Four of those bodies were exhumed in 1937 and reinterred at Pentridge Prison. A further body was exhumed from the Hospital Yard in 2003 and reburied in Fawkner Cemetery. The 2008 Pentridge Prison exhumations included the four bodies from the 1937 grave, and the Institute's analysis of those remains confirmed that they constituted four individuals, one of whom has been identified as Colin Campbell Ross who was wrongly convicted of the infamous Gun Alley Murder. His remains were released to his grandnieces following a posthumous pardon granted by the Victorian government.

One would expect that if all but five of the executed prisoners bodies buried at the Old Melbourne Goal were exhumed in 1929, then 46 bodies would have been reinterred at Pentridge Prison. However, Inspector-General Ackeroyd reported that 47 bodies were removed to the Pentridge Prison. And yet the Pentridge burial plan listed only 32 individuals in three mass graves. Some of the individuals are named and others are listed as 'unknown'. The Institute has established that in the 41 boxes or coffins that were exhumed from the three 1929 Pentridge graves, there is a minimum number of 34 individuals (meaning, of course, that it could be more), with many being incomplete or commingled (see Chapter 4).

In short, the numbers do not add up! It is quite possible that skeletal remains of prisoners executed at the Melbourne Gaol have yet to be found in the grounds of either the Old Melbourne Gaol or Pentridge Prison.

Fiona Leahy and Helen D Harris

The historical record describing the dissection of executed prisoners was supported by the dissection marks found by Institute researchers who examined the skeletal remains exhumed from Pentridge. The evidence suggests that some bodies of the executed prisoners were subjected to a full autopsy (possibly including an examination of the brain) but many were not, as less than a third of cases bore the marks of post mortem dissection. This raises the question of whether a selection process was applied to prisoners, or more limited post mortem examinations were being conducted, or perhaps some prisoners successfully exercised their right to elect not to be dissected under the School of Anatomy provisions of the *Medical Practitioners Act 1865*.

So what happened to Ned Kelly's body following his execution?

It was anticipated that Ned Kelly would have elected not to be dissected, given that he was Catholic and had made a plea to the Governor of Victoria on the eve of his execution that his body be buried in consecrated ground. And indeed, evidence of Kelly's wish was found in an article published in the *Argus* newspaper in 1891, following the execution of a prisoner called Thomas Phelan. In a discussion on whether the post mortem examination of the deceased should occur before or after the jury had viewed the body as part of the coronial inquest into the death, it was reported that:

Fig. 2.7: The death mask of Edward Feeney in the Old Melbourne Gaol. Ronn Taylor.

> He [Dr Shields] recollected distinctly having a conversation with Ned Kelly on the night preceding his execution, when the latter asked him not to touch his body after death, and he gave the requisite promise (*Argus*, 17 March 1891).

If Kelly had objected to anatomical dissection, a post mortem examination such as described in the *Bendigo Independent* was potentially unlawful. It was certainly provocative in the climate of unrest following Kelly's execution (Fig. 2.8). The police were sufficiently concerned to raise the matter with Castieau, the Governor of the Melbourne Gaol. On 25 November 1880 Superintendent Sadleir wrote to Acting Chief Commissioner Nicolson, a summary of which is preserved in the Police Register of letters and memoranda sent:

> Offender Kelly. Extract from *Ararat Advertiser* of 23rd & particulars raised in *Benalla Standard* on 23rd stating body alleg. given over after execution to Medical men etc. Unfortunate if foundation for statement as his friends will be greatly infuriated and if no foundation for statement some public contradiction of this very desirable (VPRS 676, Register of letters and memos sent Unit 25, Y1893).

Fig. 2.8: A portrait of Ned Kelly, taken the day before he was hanged. State Library of Victoria.

Nicolson wrote immediately to Castieau:

> Would Mr Castieau be so good as to inform me what truth there is in the attached para from the *Ararat Advertiser*? I quite concur with Mr Sadleir that if the statement is untrue, or if it contains a grossly exaggerated account of the disposal of the body, steps should be taken to contradict what in it is wrong (VPRS 678, Memorandum Books Unit 72, Y1893).

On 29 November the following entry is recorded:

> Govr Melb Gaol reporting no truth in attached newspaper stating that body was given to Medical men after execution and no students at examination after execution, and asks if matter wished to be made public C.Sy's [Chief Secretary's] approval be obtained first (VPRS 676, Register of letters and memos sent Unit 25, Y1911).

The same day Acting Chief Commissioner Nicolson wrote the following memorandum to Superintendent Sadleir:

> Forwarded for Mr Sadleir's information. I do not think it necessary to trouble the Chief Secretary in this matter. And in the face of Mr Castieau's memo I am

> of course debarred from making any formal contradiction to the rumour. If however a suitable opportunity should present itself there is no objection to Mr Sadleir stating how little truth there is in the paragraph. It should of course be done in an informal way and without giving unnecessary details (VPRS 678, Memorandum Books Unit 72, Y1911).

Our historical research suggested that Kelly had not been dissected after his death and that the article in the *Bendigo Independent*, re-reported in several newspapers, was not true – that medical men and students had not gone 'in heavily' and made a 'nice mess' of Kelly's body, nor souvenired body parts and his head.

Unfortunately the full memorandum from Castieau to Acting Chief Commissioner Nicolson has not been found among the boxes of police records in the Public Record Office of Victoria and it can only be hoped that one day it will surface. However, the conclusion about the lack of a post mortem examination left open the possibility that the skull given to the Institute by Tom Baxter might indeed be that of Ned Kelly!

Looking to the DNA

It was at this stage of the investigation that the DNA analysis of the Baxter skull and the Pentridge remains revealed some remarkable results (see Chapter 9).

The VIFM had located a descendant from Ned Kelly's mother Ellen Kelly (nee Quinn), who later married George King following the death of her first husband, John 'Red' Kelly. Leigh Olver was the great-great-grandson of Ellen Kelly. Although he is descended from the marriage of Ellen and George King, Leigh Olver shares the same mitochondrial DNA as Ned Kelly, as he is descended along the maternal line. He agreed to provide the Institute with a blood sample to enable DNA comparison with the skull and the Pentridge remains. The DNA analysis showed that the 'Baxter skull' was definitely not that of Ned Kelly, but allowed for his other skeletal remains to be positively identified from the remains exhumed from Pentridge.

And those remains did show distinct signs of some post mortem dissection. Only one portion of skull, a segment of the occipital bone, was among the remains and it bears vertical saw marks. This may have occurred to determine if the hanging had been effective, by looking for signs of damage to the upper cervical spine or vertebrae (see Chapter 11). Vertical saw marks in the occipital bone and upper cervical spine mean that, at least, this area was inspected post mortem, and the head at that time was attached to the body.

It is difficult to reconcile the historic record with the scientific evidence of the dissection of Kelly's body. Did Castieau mislead ACC Nicolson in denying any truth in the article published in the *Bendigo Independent*? In the 1891 *Argus* article which reported that Dr Shields gave Kelly his promise that his body would not be touched after his death, Shields added that:

> ... after the execution the sheriff said that the body was in his custody and he could instruct anyone to make a post-mortem examination.

Fact or fiction: Ned Kelly had a brother who became a policeman

Ned Kelly's half-brother, John King, did become a policeman in the Western Australian police force.

Ned Kelly had two brothers who survived him, Jim Kelly, four years younger than Ned, and half-brother John King, born to his mother's second husband, George King.

Jim Kelly was two years older than Dan Kelly, who died at Glenrowan as a part of the final shootout with police, but he was not a member of the gang as he had been arrested in 1877 for horse theft and was not released until 1880.

He and Ned's sisters visited Ned when he was in prison awaiting execution, and petitioned for his release. Later in life he became something of a recluse, first living with his mother until she died in 1923 and then living out his last years on the property of a cousin.

In 1930 he wrote a book review of J.J. Kenneally's *The Complete History of the Kelly Gang and Their Pursuers* (1929), showing that his animosity towards the police had not diminished any:

> *You have gone to no end of trouble, and displayed great patience, judgement, and tact in collecting inside official police and judicial documents and information, in order to let the world at large see for themselves how the various members of my family have been hounded down by the heads, as well as by the rank and file, of the police force.*

His obituary, as published in the *Rockhampton Morning Bulletin* of 20 December 1946, read:

> *James (Jim) Kelly, brother of Ned, the bushranger, died in his sleep early this week near Glenrowan where Ned shot it out with the police and was captured. Jim, who was over 90, was the youngest member of the Kelly family and was not involved in the bushranging life of his brothers. Years ago his memories of the period were eagerly sought by historians and other writers. He was unmarried and lived the life of a recluse with his nephew, Patrick Griffiths, on a farm between Glenrowan and Greta West.*

Jack Kelly (King) as a mounted policeman. State Library of Victoria.

John King, however, lived an altogether different life. He was born when Ned was about 20 years old and was only five when Ned was hanged.

He went by the name Jack Kelly and became a noted horseman and horse-breaker as he grew older. He and his wife Violet joined a circus as stunt riders and toured Australia, South Africa and Asia before settling in Western Australia in 1906, where he was recruited as a riding instructor for the mounted police and reached the rank of second-class constable.

He rejoined a circus a few years later, however, toured the world, became an Honorary Member of the Royal Canadian Mounted Police and served in World War I as a sergeant. In the 1920s he moved to Argentina to work as a riding instructor. He died there in 1956.

Craig Cormick

Castieau's statement to Nicolson could be interpreted as indicating that there was no dissection of the body by the medical men and their students under the School of Anatomy provisions from which Kelly had elected to 'opt out'. But that might not preclude there being a limited autopsy as directed by the Sheriff, although his legal authority to do so is dubious at best.

The Sheriff's interest in post mortem examinations was driven by an eagerness to demonstrate that the death by hanging was instantaneous, by showing a cervical fracture or dislocation (the 'hangman's fracture'). Executions were witnessed by members of the public – including journalists – and any mishaps were reported in the papers the following day. And there were such mishaps, due to the rope being too long or too short or the noose being obstructed by clothing. Bungled hangings damaged the reputation of the Sheriff and undermined the authority of the state: Dr Shields or another medical practitioner may have been directed by the Sheriff to examine Kelly's cervical vertebrae to ensure the hanging had been effective (see Chapter 11). Evidence of these types of post mortem examinations can be found in the Sheriff's record of executions from 1894 to 1967, called the Particulars of Executions.

Unanswered questions

In addition to seeking identification of the skull handed in by Tom Baxter (which was matched with a case from the Pentridge remains which is both incomplete and commingled) is the enduring question of the location of the remainder of Ned Kelly's skull. There are many theories. Most people are still wedded to the notion that the skull was taken by the dissecting doctors but it is more than possible that it was souvenired in 1929, when the graves at the Melbourne Gaol were ransacked by onlookers.

If the skull is ever discovered it will be easily identifiable, as well as distinguishable from false claims, as it will be incomplete with a portion of the occipital bone missing. It would be a very welcome conclusion to our investigations if the remainder of Kelly's skull could be found and buried with his body, now resting in the Greta Cemetery in north-east Victoria.

Fact or fiction: No one knows when Ned Kelly was born

Ned Kelly was the third of 12 children born to Ellen Kelly (from three different fathers), but there is no clear evidence of his actual birth date. He was most likely born somewhere between the end of 1854 and mid 1855, near Beveridge just north of Melbourne. Unlike his siblings there is no record of his baptism either, so Ned's age at his death is a little uncertain. He may have been either 25 or 26 years old when he died.

According to the *Australian Dictionary of Biography*, Ned's mother Ellen Kelly was the fourth of 11 children. She arrived in Port Phillip as an assisted migrant from County Antrim in Ireland in July 1841. She took up with a former convict, John 'Red' Kelly, who had been transported to Van Diemen's Land for theft, and fell pregnant to him in May 1850 at the age of 18. They married in November of that year and their first-born child, Mary Jane, died while young. They then had a daughter Anne, followed by Ned, named after John's brother Edward.

In 1866 John Kelly died, leaving Ellen with eight children. Ellen later had another child to Bill Frost before marrying George King, a 23-year-old Californian horse-thief, who also got her pregnant. She had three children to King, who subsequently abandoned her.

She was in the Old Melbourne Gaol serving three years for the attempted murder of Constable Fitzpatrick when Ned was hanged there in 1880. After her release she raised three of her grandchildren after her daughters Maggie and Kate died in the late 1890s. She died in 1923, aged about 91; of her 12 children, one son and daughter of her first marriage and a son and two daughters of her second marriage survived her.

Craig Cormick

References

Bendigo Independent, 19 November 1880.

'Colonial recollections'. *South Australian Chronicle and Weekly Mail*, 15 December 1877.

Criminal Law and Practice Statute 1864.

Finnane M (ed.) (2004) *The Difficulties of My Position: The Diaries of Prison Governor John Buckley Castieau, 1855–1884*. National Library of Australia, Canberra.

'Help needed in mysterious case of Ned's head'. ABC/AAP, 20 June 2010.

http://www.austlii.edu.au/au/legis/vic/hist_act/

Jones I (1995) *Ned Kelly: A Short Life*. Lothian Books, Melbourne.

MacKinnon D (2012) Bodies of evidence: dissecting madness in colonial Victoria (Australia). In: *The Body Divided: Human Beings and Human "Material" in Modern Medical History*. (Eds S Ferber, S Wilde) pp. 75–108. Ashgate.

Medical Practitioners Act 1865.

National Trust correspondence, January 1969.

'Prison curiosities'. *Argus*, 12 May 1894.

Public Record Office Victoria VPRS 24/P0, Unit 411, Inquest No. 938 of 1880, *Edward Kelly*.

Public Record Office Victoria VPRS 676/P0, Unit 25, Register of letters and memos sent.

Public Record Office Victoria VPRS 678/P0, Unit 72, Memorandum books.

Public Record Office Victoria VPRS 4965/P0 Part 1 Police Branch Unit 2 Item 26: From ACP Nicolson re: Execution of outlaw Edward Kelly extra police required in district.

Public Record Office Victoria VPRS 4966/P0 Kelly Historical Collection – Part II Crown Law Department.

Sadleir J (1913) *Recollections of a Victorian Police Officer*. George Robertson and Co., Melbourne. Reprinted 1973, Penguin Books.
'State probes Ned riddle'. *Herald Sun*, 13 November 2009.
Sunday Herald Sun, 9 March 1997.
Victorian Parliamentary Debates (1886) Legislative Council, 29 June, 464.

Chapter 3
Analysing the skull

Soren Blau and Chris Briggs

Initial examinations of the skull alleged to be Ned Kelly's were conducted by forensic anthropologists and pathologists. They found it was of the right age, sex and ancestry to be Ned's skull and that some damage had occurred to it over the years, and that it had been crudely mended.

When the skull that was believed to be Ned Kelly's was handed into the Victorian Institute of Forensic Medicine by Western Australian farmer Tom Baxter, it was treated in accordance with processes that govern all remains admitted to the Institute. The skull, which became known as the 'Baxter skull', was first given the case number 5334/09, before being CT-scanned and examined by a forensic anthropologist and forensic pathologist.

Forensic anthropology is the field of study concerned with the examination of material believed to be human, primarily to answer medico-legal questions, including attempts to identify the remains (Blau and Ubelaker 2009; Dirkmaat 2012). Skeletal remains are admitted to the Institute because they are 'reportable deaths' under the *Coroner's Act*. They are then examined by a forensic anthropologist, whose job is to:

- provide a written and photographic description and record of the condition and preservation of the remains. The level of preservation will influence the detail to which other analyses can be undertaken;
- determine, where possible, the forensic significance of the skeletal remains. While each state and territory in Australia has slightly different legislation on when human skeletal remains can be legally considered of 'forensic significance' (Blau 2009), remains more than 50–100 years since death are, in most circumstances, considered of no medical or legal significance (due to the fact that if the person was the victim of a homicide, the likelihood of the perpetrator being alive is virtually nil);
- estimate the minimum number of individuals represented by the skeletal remains. This may be important in cases where remains may have become commingled or mixed, with one or more other sets of remains;
- provide a biological profile which includes information about the individual's ancestry, sex, age and stature when they were alive;
- describe and analyse any disease (pathology) and/or injuries evident on the skeletal remains. All incidences of skeletal pathology and trauma are described in terms of the location on the body and physical impact. The forensic anthropologist may also form an opinion about the timing of any skeletal changes. These could include injuries or disease that occurred while the individual was still alive (ante mortem), or that

occurred around the time of death (peri mortem). Damage from animals, natural processes or human activity that occurred after death (post mortem), which can include damage during the recovery process is also recorded.

When fragmentary or damaged skeletal remains are examined, the forensic anthropologist may use a hot melt glue gun, super glue or dental wax to attempt to put them back together. Such reconstruction assists the forensic anthropologist and pathologist in their interpretation of the skeletal damage.

In the case of the 'Baxter skull', the anthropologist concluded that it came from an adult Caucasoid male. While predominantly intact, the skull had some damage including a thin vertical cut (about 2.5 mm wide × 11 mm long) through the body of the left lower jaw (mandible), as well as evidence of attempted repairs, including the presence of brown wax, orangey glue and even Bluetac adherent to some of the bony surfaces.

The presence of a partial metal hook and spring wire inserted in the right part of the lower jaw also suggested that at some point the jaw had been artificially attached to the skull. All of the upper teeth had been lost sometime after death, but some of the lower teeth were still present in the jaw. The skull showed no evidence of any skeletal disease or trauma, but the words 'E. Kelly' had been hand-written in black ink on the left side of the cranium (on the temporal bone) (see Fig. 3.1).

While an anthropological examination alone was not enough to determine the time since death, the evidence of damage and repair indicated that the skull was quite old. As the

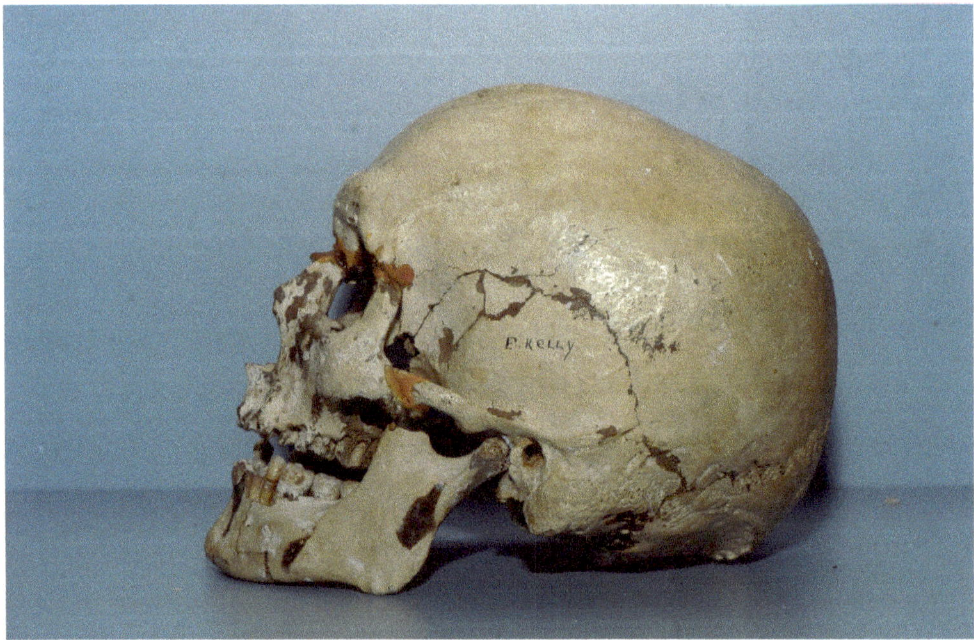

Fig. 3.1: A view of the left-hand side of the 'Baxter skull', clearly showing the words 'E. Kelly' hand-written in black ink. Victorian Institute of Forensic Medicine.

initial analysis had found nothing to indicate that the skull might not be that of Ned Kelly, it was passed along to the next stage of examination, looking in detail at the teeth and CT scans.

Note: the majority of this chapter is based upon S. Blau *et al.* (2014) The contributions of anthropology and mitochondrial DNA analysis to the identification of the human skeletal remains of the Australian outlaw Edward 'Ned' Kelly. *Forensic Science International* **240**, e11–e21.

References

Blau S (2009) Forensic anthropology and police investigations. *Australian Police Journal* **63**(1), 8–14.

Blau S, Ubelaker DH (eds) (2009) *Handbook of Forensic Anthropology and Archaeology*. Left Coast Press, California, pp. 457–467.

Dirkmaat D (ed.) (2012) *A Companion to Forensic Anthropology*. Wiley-Blackwell.

Who does what at the Victorian Institute of Forensic Medicine

The identification of Ned Kelly's remains required the expertise of a group of forensic specialists including pathologists, odontologists, anthropologists, molecular biologists and radiologists. Two people from each of these disciplines worked together, bringing their specific areas of expertise to the task of identification.

Forensic pathology

The role of the forensic pathologist is to assist coroners, police and courts to resolve critical medical issues, including causes of death, the circumstances in which deaths occurred and how injuries might have been caused.

Forensic pathologists focus on the examination of deceased people. This examination includes a review of medical histories, external examination and CT scans. Sometimes internal examinations or autopsies are performed to help resolve the issues. In this work, the forensic pathologist is assisted by technical staff in a range of further investigations, such as toxicology, histopathology, DNA testing and microbiology.

Forensic odontology

Forensic odontology is a subdiscipline of dental science which involves the relationship between dentistry and the law. Forensic odontologists are forensic dentists.

Odontologists, working in a variety of jurisdictions, produce reports and give evidence in the Coroners' Courts, the Supreme Court, the County Court and the Magistrates Court.

They are asked to provide their services in a variety of circumstances involving the identification of the deceased, facial trauma analysis, the analysis of bite marks, facial approximation/reconstruction, superimposition and disaster victim identification.

Forensic anthropology

Forensic anthropology is the examination and analysis of a range of skeletonised human remains. This may include the recovery and scene interpretation of buried and/or surface scatters providing evidentiary and investigative evidence.

The forensic anthropologist has a sound grounding not only in human skeletal anatomy but also in the effects of trauma, time, burial and other taphonomic factors upon skeletal remains.

Molecular biology

The VIFM's Molecular Biology Laboratory (MBL) provides DNA profiling services for the identification of human remains, missing persons investigations and other matters that require DNA analysis.

In Ned Kelly's case the team relied on mitochondrial DNA analysis.

Mitochondrial DNA (mtDNA) is a form of DNA that can be recovered and analysed to yield what are known as mtDNA profiles (or haplotypes). Mitochondrial DNA is present in greater numbers in each cell (100–1000s copies per cell) than nuclear DNA (nDNA) (one copy per cell), and is fairly robust in nature. Unlike nDNA, mtDNA is maternally inherited; hence, all the individuals in the maternal lineage in a given family will share the same mtDNA profile.

Forensic radiology

In 2005 a computed tomography (CT) scanner was installed in the mortuary of the VIFM. It was one of the first mortuary-based CT scanners in the world and since that time, all deceased persons admitted to the Institute have been CT scanned and the images permanently stored on an archiving system. As a result, the VIFM is a world leader in the application of sophisticated CT imaging in forensic practice.

The CT scanner does not replace the forensic autopsy but can provide valuable assistance to pathologists in preparing for autopsy and in documenting injuries, images of which can later be presented in courts. CT has provided extremely valuable assistance in identification of the deceased in a mass disaster situation, as was the case in the 2009 'Black Saturday' Victorian bushfires.

Deb Withers

Chapter 4
Bringing up the bodies: the search for the lost Pentridge burial ground

Jeremy Smith

The discovery and subsequent recovery of bodies at Pentridge Prison was a complicated task – made more difficult by the fact that no one seemed to know where the bodies might be. They were eventually all found, by archaeological and historical detective work. It was only then that a full excavation could be undertaken. And it revealed some very interesting findings. The interest in the bodies also prompted Tom Baxter, a man with a skull and a story to tell, to come forward.

In March 2009, at the bottom of a deep pit at the former Pentridge Prison, a team of archaeologists from Heritage Victoria literally lifted the lid of the makeshift coffin that, although unknown at that point, contained the remains of Ned Kelly. The discovery brought to an end a program of archaeological and historical investigations that had started six years earlier.

The exhumation of the remains, and their delivery to the Victorian Institute of Forensic Medicine (VIFM), enabled the next step in this unique project to begin: the forensic investigations that would prove definitively that the remains of Ned Kelly had been found. The project has been a landmark example of how the disciplines of archaeology, history and forensic science can work together to achieve results which would not be attainable in one field of expertise alone.

At the start, the project was not even about the search for Kelly. Following archaeological work at the Old Melbourne Gaol in 2002, it had become clear that our understanding of the burial locations of many of Melbourne's executed prisoners was not at all accurate. In fact, with the benefit of hindsight, I can now look back to 2002 and recognise that almost everything that we thought we knew at that time about prisoner burial locations was incorrect. So, initially, my aim as Heritage Victoria's Senior Archaeologist was to make sense of the results emerging from the work at the Old Melbourne Gaol, and ensure that the location of all prisoner burials was known beyond doubt. This information was needed because many of the bodies from the Old Melbourne Gaol had been moved to Pentridge in the late 1920s. And more recently Pentridge has gradually undergone a transformation from gaol to urban development zone, following its closure and sale to property developers in the late 1990s.

It may seem surprising that in 2002 the precise burial location of so many people was in question. Forty-three men and women were executed at the Old Melbourne Gaol and Pentridge Prison between 1880 and 1967, including significant (if notorious) historical figures such as Frederick Deeming (see p. 117), poisoner Martha Needle, the 'baby farmer' Frances Knorr, Jean Lee (the last woman executed in Australia), Ronald Ryan (the last person executed in Australia) and of course Ned Kelly.

The lack of certainty was due in part to the treatment that was given to the remains of executed inmates, for scant regard was given to the condition or recording of their burials. Behind the bluestone prison walls of Pentridge and the Old Melbourne Gaol, graves were not well marked and were sometimes disturbed or moved without notation in official records. Knowledge of burial locations seems to have been lost one or two decades after the initial interments took place. For example, in 1937 authorities were unable to relocate all of the five burials that had taken place at the Old Melbourne Gaol between 1916 and 1924.

But without doubt the most significant event in the history of Melbourne's prisoner burials (and the reason for the uncertainty that surrounded the details of the Pentridge burials) took place in April 1929, when the Labour Yard that also served as a burial ground at the Old Melbourne Gaol was dug up, and souvenir hunters rushed in to grab bones. The bones that escaped the attention of trophy collectors were boxed and bagged, and relocated to Pentridge. The scene of chaos was described by the *Argus* as:

> The edge of the shovel ripped off the lid of the coffin, which was made of redgum, and a skeleton was revealed. As soon as this gruesome discovery was made, a crowd of young boys who had been standing round expectantly while eating their luncheons rushed forward and seized the bones. The workmen intervened, but in the scramble portions of the skeleton were carried off, including even the teeth (*Argus*, 13 April 1929).

There was a public outcry as details of the works at the gaol became known. In response a professional undertaker, Josiah Holdsworth, was engaged to oversee the clearing of the rest of the former burial ground, and the transfer of some remains to new coffins. However, it is clear that by the time of Holdsworth's involvement many of the gaol burials had been disturbed, and were perhaps already on their way to Melbourne's other main prison, Pentridge, for reburial in unmarked mass graves.

Also of interest was the fate of the gaol bluestone wall that many prisoners' initials were carved on (see Fig. 4.1). Many of the bluestones were salvaged and set aside for reuse; interestingly, some of them were then used in the construction of the sea wall that runs along the eastern side of Port Phillip Bay, through the bayside suburbs of Brighton, Beaumaris and Black Rock. This sea wall was built as part of Depression-era civic works in the late 1920s and 1930s.

Today, if you wander along the wall, there are three places where you can see clusters of stones carved with initials and dates, such as 'WC 24.8.91', 'AA 21.11.98' and 'C.H.S 13.1.96' (marking the burials of William Colston, executed 24 August 1891; Alfred Archer, 21 November 1898; and Charles H. Strange, 13 January 1896). A total of 18 stones carved with prisoner initials and execution dates have so far been rediscovered along the sea wall, but undetected ones may survive built into the foundation of the wall or facing into the wall's interior. The current location of the stone bearing the 'EK' initials (and the broad arrow

Fig. 4.1: The partial demolition of the Old Melbourne Gaol in 1929. Some of the inscriptions of prisoner's names can be seen in the historic photograph. They are visible as a row of small light-coloured squares just above the base of the wall. The photo shows the demolition of the southern wall of the Labour Yard. The northern wall remained standing until 1937. Image courtesy of RMIT University.

mark) is not known – it may not have survived the demolition works at the gaol. A stone carved with the details 'E.K 19.3.94' is on display in one of the cells at the Old Melbourne Gaol – but this is the marker of Ernest Knox (not Kelly), executed in March 1894.

The Pentridge reburials

In late April 1929, as the stones from the wall were being sent south to Melbourne's bayside suburbs, the exhumed bones that had escaped the attention of ghoulish trophy-hunters went north to Pentridge for reburial. Unfortunately the location of the burial ground used for the reburials at Pentridge was never formally documented. The same area was used nine times in the 20th century, to bury inmates executed at Pentridge between 1924 and 1951. But when Ronald Ryan was executed in 1967, a different area was chosen for his burial and it seems that knowledge of the location of the earlier burial ground was starting to fade from memory.

In fact, when Pentridge Prison was closed and then sold by the state government to private developers in 1997, it was widely believed that all of the burials (including the remains relocated from the Old Melbourne Gaol, and the 10 prisoners who were executed and buried at Pentridge in the 20th century) were buried in a triangular plot of land adjacent to the east end of the large bluestone D Division building, where former wardens remember Ryan being buried. From the time of Ryan's burial onwards, this area came to be known as the 'Pentridge Cemetery'. Sometime after Ryan's burial, a basic metal and bluestone monument to Ned Kelly was erected in this area, supposedly marking Kelly's burial place.

Fortunately, the archaeological project conducted by La Trobe University at the Old Melbourne Gaol in 2002 brought the question of the Pentridge burials into focus again, and flagged the need for Heritage Victoria to coordinate further investigations.

Surprising discoveries

In 2002, RMIT approached Heritage Victoria for permission for building and landscaping work in the grounds of the former Melbourne Gaol hospital. We recognised the potential for the site to contain significant archaeological remains, and required RMIT to engage project archaeologists. RMIT employed a team from the Department of Archaeology at La Trobe University, who made a series of surprising discoveries as they unearthed the archaeology of the Old Melbourne Gaol (Wright 2002; Hewitt 2003; Hewitt and Wright 2003; Zygmuntowicz 1998).

One of the first was a pit containing the intact skeletons of six dogs (see Fig. 4.2). Initial thoughts were that they may have been the remains of gaol guard dogs, but there is no evidence that dogs were ever used at the gaol. In a remarkable example of archaeological evidence and historical research being used to make sense of a puzzling situation, the project team tracked down an article in *The Australasian Sketcher* from 1877, titled 'Sketches of snakebite experiments in the Melbourne Gaol' (see Fig. 4.3). The article shows that at that time the Victorian Snake Bite Commission was conducting (clearly unsuccessful) experiments on tiger snake antivenom, using dogs at the gaol.

Soon after the discovery of the pit containing the dog skeletons, the La Trobe University archaeologists made a more significant discovery. They uncovered the remains of a largely intact coffin against the southern wall of the former gaol hospital compound. Research revealed that the grounds of the hospital had been used for five prisoner burials between c.1916 and 1924. In 1937, four of these interments had been exhumed and relocated to Pentridge, but the fifth burial could not be found at that time. The intact coffin discovered by the La Trobe archaeologists was the missing burial that the 1937 team had been unable to find.

Another important discovery made at this time was a historical document, not an archaeological find. As part of the background research for the project, the La Trobe team acquired copies of a couple of different versions of a plan that seemed to show the layout of a burial ground at Pentridge. One version of the plan shows an area, rectangular in shape (with marked dimensions of 80 feet × 24 feet) containing several burial plots (see Fig. 4.4). Some of these are the graves of individuals; others are mass graves containing multiple sets of remains. The names of the individuals purportedly located within each burial are marked on the plan, beneath each plot. And Ned Kelly's name is marked beneath the largest of the mass graves.

Another burial ground?

There are two things of particular significance about the burial plan. The first is that the burial place and name of Ronald Ryan are not shown, suggesting that the plan predates 1967. The second is that the burial area is of a totally different size and shape from the so-called Pentridge Cemetery area, located near the D Division building, which is triangular. In short, the plan suggested that there may have been an earlier burial ground at Pentridge,

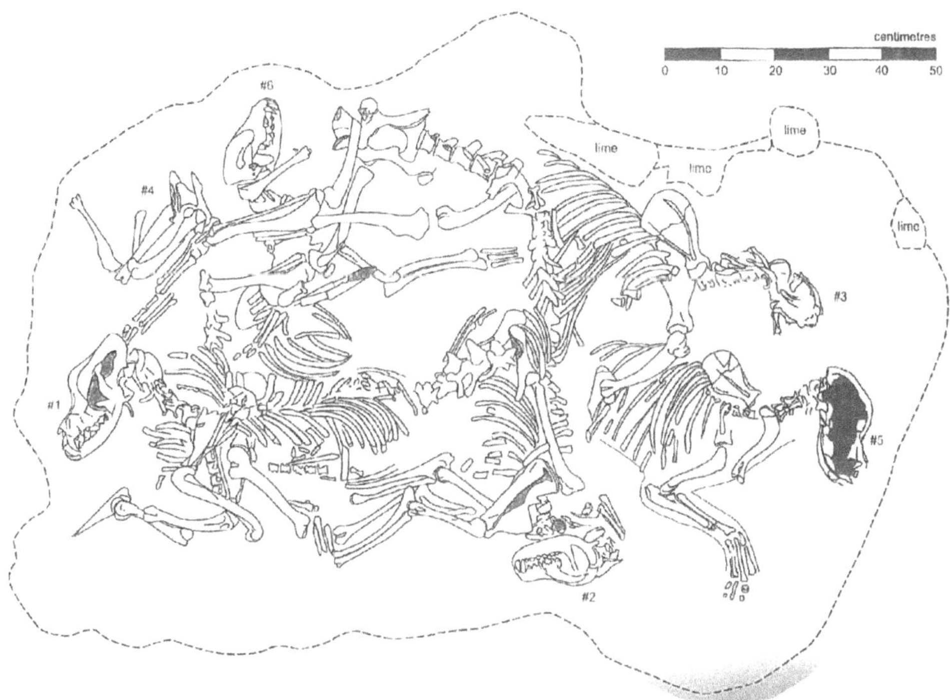

Fig. 4.2: Plan of dog burials, Old Melbourne gaol, 2002. Plan courtesy of Geoff Hewitt, Department of Archaeology, La Trobe University.

Fig. 4.3: *Australasian Sketcher* article, February 1877.

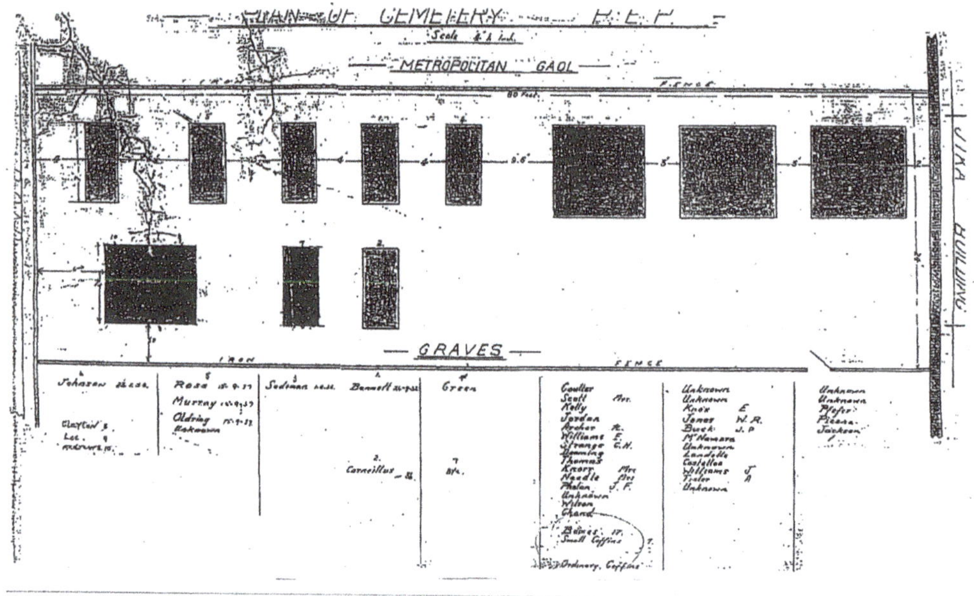

'Plan of Cemetary P.E.P. Metropolitan Goal' (undated)

Fig. 4.4: Pentridge Prison burial plan (undated).

different from the one where Ryan had been buried in 1967 and in another – and unknown – location.

As a result of the questions raised by the burial plan and as development of the Pentridge site gathered speed, I recommended to the site owner, Pentridge Village Pty Ltd, that archaeological testing take place in the Pentridge Cemetery. The testing was needed to clarify whether all of the remains relocated from the Old Melbourne Gaol (about 33) and the remains of the 10 prisoners executed at Pentridge between 1924 and 1967 could be accounted for. In 2007 archaeologists from heritage consultancy TerraCulture, directed by Catherine Tucker, opened the area for excavation. The results of the investigations were conclusive – the so-called Pentridge Cemetery contained only one burial, that of Ronald Ryan.

So where were the other bodies? The remains of some of the state's most notorious individuals, including perhaps the most famous of all Australians, were in an unknown burial location, somewhere on a development site.

Turning to the records

In response, we immediately commissioned two historians, Helen Harris and Carlotta Kellaway, to search prison records, looking for details of the historic burial ground. A significant breakthrough came with discovery of the 'Airpsy' series of aerial photography, from the Pictures Collection of the State Library of Victoria, which showed the Pentridge site on 25 May 1955 (see Fig. 4.5). One image, when magnified, showed a rectangular yard, unused and covered with tall weeds, enclosed by a corrugated iron fence (see Fig. 4.6). The dimensions and shape of

Fig. 4.5: 'Airspy' image of Pentridge Prison, February 1955.

Fig. 4.6: 'Airspy' image, February 1955 (close-up).

the yard matched the area shown in the prison records plan. In March 2008 the TerraCulture archaeologists were again engaged to conduct testing in this area.

After a few days the TerraCulture team discovered that the area shown on the 1955 aerial photo was indeed the location of the historic burial area. They uncovered the burials of all nine inmates executed at Pentridge between 1924 and 1951, all with nameplates still on their coffins (see Fig. 4.7). They also found a small mass grave containing the four inmates relocated from the Old Melbourne Gaol in 1937, and two of the three large mass graves containing prisoner remains relocated from the gaol in 1929. One of the mass graves contained 12 coffins, the other contained five. A noticeable feature of the burials was the use of large quantities of lime inside the coffins and packed between them, presumably intended to accelerate the deterioration of the remains. Ironically, most of the lime had hardened and set like plaster, preserving the remains and having the opposite effect from what was intended. In accordance with Victorian legislation, the remains that did not have identification plates were exhumed and delivered to the coroner.

Disappointingly, the mass grave shown on the burial plan containing the largest number of burials (including Kelly) was not located in the area shown on the plan, and could not be found by the TerraCulture archaeologists. What they did find was that this part of the site had been extensively disturbed by the construction of a storm drain, perhaps in the 1960s; they considered it possible that the missing mass grave may have been disturbed and removed

Fig. 4.7: One of the single burials identified by the archaeologists in the burial ground. Heritage Victoria.

at that time. There were anecdotal reports from former Pentridge warders that around the 1960s some bones had been disturbed during works at Pentridge. Those bones were discarded in the nearby quarry, which today is located under a new housing estate. As a result, around the middle of 2008, the media were reporting the likelihood that Ned Kelly's remains would never be found.

In fact, the extensive media coverage and speculation led to an unexpected turn of events that was to prove crucial to the ultimate success of the project. In mid 2008 I received a phone call from Mr Tom Baxter. I had heard of Tom Baxter before – he was the person associated with the theft of a skull from the Old Melbourne Gaol in 1978. The skull, which was on display at the gaol at the time, was one of four that had been given to the National Trust by the Institute of Anatomy in Canberra in 1971. The skulls were believed to be those of executed inmates and, at the time of the theft, one of the skulls was being displayed as that of Ned Kelly, alongside his death-mask. The skull had the name 'E. Kelly' written on the left side of the cranium. It is likely that these four skulls were collected from the gaol yard during the site disturbance and exhumations in 1929 by Mr Harry Lee, whose building firm of Lee and Dunn was employed on site.

A few weeks after the disappearance of the skull from the gaol in 1978, Tom Baxter had come forward offering to return the skull if it could be buried in consecrated ground outside gaol property. Of course in the 1970s it was not possible to determine whether the 'Baxter skull' was in fact Ned Kelly's, and not surprisingly no deals were done. For the next 30 years, Tom Baxter remained in possession of the skull, which was hidden on his property in the remote Kimberley region of Western Australia, despite occasional unsuccessful attempts by the police to recover it.

Throughout 2008 and 2009, Tom regularly contacted me and asked for updates on the discoveries at Pentridge. He was interested in whether any progress had been made in the search for Kelly. Whenever we spoke, I encouraged him to hand in the skull and emphasised that its return might assist the forensic investigations that could give conclusive answers on the identity of the Pentridge remains. I also knew that the skull was an object of intense interest – it was needed to generate the public and political support that would facilitate the forensic work that followed the archaeological discoveries. Despite the discovery of the mass graves at Pentridge, the archaeologists at Heritage Victoria and the forensic experts at VIFM knew that any attempt to establish the identities of the prisoner remains would be complex and time-consuming, and lay outside the core responsibilities of both agencies.

The Glenrowan connection

Fortunately, in November 2008, Tony Robinson of the UK Channel 4's *Time Team* fame came to Glenrowan to film a documentary on the archaeology of the Ann Jones Inn, site of the Kelly Gang siege in 1880. The site had already been excavated by archaeologist Adam Ford (see Chapter 18) of Dig International, over a six-week period in April and May of 2008 as part of the Glenrowan Revitalisation project, and most of the *Time Team* documentary was based on Adam's findings. I invited Tom Baxter to visit Glenrowan during the filming and suggested that this might also be a good time for him to return the skull.

I didn't really expect Tom to turn up at Glenrowan, so I was a little surprised when I noticed him during a break in the filming, standing against a fence. I hadn't met him before, but recognised him from newspaper articles and interviews that he had done in the 30 years since the skull's disappearance. We spent the next couple of hours walking around the archaeological trenches that were being excavated, looking at some of the artefacts that had been found and discussing all things Kelly. I learnt that for many years the skull had been sealed inside a plastic Tupperware container, hidden in an old tree stump on his remote property. Sometimes the area flooded, and the container bobbed around in the water for months at a time.

He didn't give me the skull at our Glenrowan meeting and I don't think he had it with him, but he did give me a tooth which he said came from the skull. He also gave me a report he had commissioned from the School of Anatomy and Human Biology at the University of Western Australia, which included an assessment of the skull as being that of an adult male, 20–30 years of age. I left Glenrowan without the skull, but Tom and I continued to talk regularly in late 2008 and early 2009.

Finding the missing bodies

But the most significant discovery of all was yet to come. In February 2009 I was contacted by the owners of the Pentridge site and told that something of potential interest had been uncovered. An object that looked like part of a wooden box had been exposed at the bottom of a newly excavated service trench, not far from the historic burial ground. I immediately headed out to inspect the discovery, entered the pit and cleared around the partially exposed box with my trowel. As the top of box was exposed, it became clear that several other boxes were also buried close by. But these were boxes, not coffins. My suspicion that the missing mass grave had been found was tempered by the obvious differences between this feature and the other mass graves, which had contained proper coffins not smaller boxes. However, all doubts were dispelled when I lifted the lid of the box, and exposed a partial set of human remains. After urgent discussions with the coroner's office and VIFM it was agreed that Heritage Victoria's archaeology team would conduct the exhumation of all burials located in the newly discovered mass grave.

The next six weeks were the most challenging and most satisfying time of my career in archaeology, as it became clear that we had found the large mass grave shown on the historic burial plan, albeit in a slightly different location. Each day, the team worked in the mass grave, exposing individual burial boxes and carefully recording their positions and other significant details. Rather than conduct actual exhumation work in the open area at Pentridge, we relocated the burial boxes, one at a time, to Heritage Victoria's archaeology laboratory in Abbotsford.

Each evening, after a day's work at Pentridge, members of the team met at the laboratory and opened a single burial box. The human remains were lightly cleaned, recorded and packed for delivery to the coroner. Sometimes other items found in the boxes threw light on the circumstances of the burials; one box contained fragments of the *Argus* newspaper from April 1929, the time of the relocation from the Old Melbourne Gaol to Pentridge – perhaps rubbish discarded in the burial box along with the bones, by one of the demolition team. Four of the boxes had the name of an inmate and the date of his execution scrawled in faint lead pencil on their lids – perhaps an attempt by a workman to preserve the identity of the prisoner whose remains were being moved.

As our investigation of the mass grave continued, two points of particular interest emerged. First, the mass grave (which was almost 2 × 2 m in dimension) consisted of two layers of burials, one directly beneath the other (see Figs 4.8 and 4.9). There were 15 burial boxes on the top layer, and nine underneath.

The second point was that, unlike the other Pentridge mass graves which contained only coffins, this one contained seven coffins but also 17 wooden boxes of various shapes and sizes. In all likelihood, these boxes had come from the Melbourne Gaol, where they had been used in different ways before being repurposed as burial boxes. The stencil on one box, for instance, showed that it was once a storage crate for lamp kerosene from the Commonwealth Oil Refineries. Four identical long narrow boxes (including the one which was later found to contain Ned Kelly's remains) were manufactured by the Mann Edge Tool Co. in Pennsylvania, and originally contained axe or sledgehammer handles (see Fig. 4.10). A close examination of the bottom left-hand corner of the 1929 image (see Fig. 4.1) shows the demolition works at the Melbourne Gaol, with several boxes stacked in a pile. These were the same boxes that we unearthed 80 years later in the mass grave at Pentridge.

Fig. 4.8: The top layer of the mass grave at Pentridge Prison, March 2009. Heritage Victoria.

Fig. 4.9: The bottom layer of the mass grave at Pentridge Prison, March 2009. Heritage Victoria.

The reuse of boxes, rather than coffins, for the reburial of some of the executed inmates hints at a general disregard for the condition of the prisoner remains. The differences noted between this mass grave and the two investigated by TerraCulture in 2008 (which contained only proper coffins) may be explained by the sequence of events that took place in April 1929. Ironically, even though the mass grave investigated by Heritage Victoria in 2009 was the last feature to be found by the archaeologists at Pentridge, it probably represents the first phase of the relocations undertaken in 1929. The use of the utilitarian boxes was the work of the construction team, responsible for the chaotic conditions reported so vividly in the *Argus* around 12 April 1929, before the engagement of the undertaker Josiah Holdsworth a few days later. Under Holdsworth's supervision it is likely that new coffins were used, if required. The mass graves investigated in 2008 by TerraCulture were the work of Holdsworth.

As our exhumation and recording work at Pentridge drew to its conclusion, it became clear that the degree of preservation of the individual remains within each box or coffin varied considerably. Most of the coffins had a high level of intactness, with many of the bones having survived. Some of these coffins may have contained the remains of inmates executed at the Old Melbourne Gaol in the early 20th century and, as the coffins were not old when the relocation took place in 1929, they may still have been in good enough condition to be moved across to Pentridge without even being reopened. Conversely, many

Fig. 4.10: The 'Mann Edge Tool Co.' box used as a coffin for Ned Kelly's remains, that originally contained axe or sledgehammer handles. Heritage Victoria.

of the burial boxes contained fewer bones and more fragmentary sets of remains. Some commingling of remains across different sets of skeletons was also evident in some burials.

We knew that the mass grave that we were investigating was the one that the historic burial plan identified as containing the remains of Ned Kelly. An annotation on the burial plan for the mass grave shown as containing Kelly read 'Boxes: 17 Small Coffins: 17' (see Fig. 4.4) and this correlated exactly with the number and type of boxes and coffins that we had found. However, I thought that if any definitive identification of Kelly could be made, it would be likely that only a few of his bones had survived. I was mindful of the newspaper reports from 1929 which talked about the collection of bones as trophies, and the targeting of Kelly's marked burial in particular.

Security needed

The high level of media interest associated with the project meant that 24-hour security was required to protect the grave sites from disturbance by members of the public. On one occasion, the guard caught a visitor attempting to scale a site fence in the middle of the night. The individual was a passionate Ned Kelly enthusiast, who stated that he just wanted to visit the burial and souvenir a rock from the site. The construction workers employed at Pentridge also took a strong interest in our work, coming over on their breaks to check on our progress and offer advice. And it was one of the workers who, having observed the work of the TerraCulture archaeology team in the main burial area in 2008 and been instructed about the need to report potentially significant finds, realised that he had found something when the burial box from the missing mass grave was exposed and reported to me during construction works in early 2009.

I think the discovery of the mass grave that was likely to contain the Kelly remains may finally have convinced Tom Baxter to change his mind about his custodianship of the skull. In mid 2009, he called me and said that he had made the decision to hand it in. I contacted VIFM and the Coroner's Office, and arrangements were made for delivery of the skull on the date chosen by Tom, the anniversary of Kelly's execution on 11 November. He never asked for indemnity from prosecution, although he never said that he was responsible for its theft (and I never asked him that question).

However Tom is regarded, he is certainly passionate about the legacy of Kelly and I am grateful that he made the decision to return the skull, as without its return I believe that the definitive identification of Kelly's skeleton may not have been possible.

The skull's return generated a high level of public interest, and the VIFM was then given the political support and resources needed to examine both the skull and the rest of the Pentridge remains in detail. Until that time, it had not been clear to Heritage Victoria or the VIFM whether it would be possible to conduct the complex investigations needed to identify the individual prisoners, including Ned Kelly. There was a real possibility that once the coroner has completed her determination about whether inquests were needed, the remains would be simply reburied at Pentridge in a designated area unaffected by the new development works. The opportunity to make definitive identifications would have been lost. The return of the 'Baxter skull' changed this, to my mind, and enabled the VIFM to launch into its remarkable program of investigations and discoveries.

Final reburials

In many ways the project did not end for me until December 2012, when I returned to Pentridge with a couple of team members. Our task was to rebury the remains of the prisoners, although this time they were each reinterred in new burial boxes in individual plots rather than mass graves (see Fig. 4.10). They were buried in a large area adjacent to the east end of the D Division building, including the area of the triangular 'Pentridge Cemetery' where Ronald Ryan had been buried in 1967. I took particular care to record the precise location of the burials, and forwarded the details to a range of interested parties and officials to ensure that the information is never again lost!

No development works will be allowed in this area. It will be fenced and lightly landscaped, and some interpretative heritage signage will be installed to present the history of the area and the people who are buried there. The 'Baxter skull' is buried in this area too, following a DNA match between the skull and a set of remains that proved not to be Ned Kelly's – they have now been reunited.

Ned Kelly's remains are, of course, not in this 'new' burial area at Pentridge. After being definitively identified, they were returned to his family in January 2013, to be buried in an unmarked grave alongside other family members in a private ceremony in Greta in northern Victoria. It is, by my reckoning, the fourth time Ned Kelly has been buried. I hope his remains will now lie undisturbed.

References

Ned Kelly's grave. *Argus*, 13 April 1929.

Record of murderers. *Argus*, Friday 4 June 1937.

Hewitt G (2003) *Archaeological Investigation at the Former Police Garage Site, Russell Street, Melbourne. Final Report, Part 1.* Department of Archaeology, La Trobe University.

Hewitt G, Wright R (2004) Identification and historical truth: the Russell Street Police Garage burials. *Australasian Historical Archaeology* **22**, 57–70.

Wright R (2002) Archaeological investigations at the former Police Garage site, Russell Street, Melbourne, Report on Burial: Designated F85 from the former Police Garages on Russell Street, Melbourne. Report to Heritage Victoria and RMIT University Property Services, unpub.

Zygmuntowicz K (1998) Old Melbourne Gaol Hospital and Warders' Yards: research on the history of the Police Garage site, Russell Street, Melbourne. Prepared for RMIT Asset Management Group and Department of Archaeology, La Trobe University (May).

Chapter 5
Anthropology: identifying the skeleton by its injuries

Soren Blau and Chris Briggs

While a large number of complete and incomplete skeletal remains were recovered from Pentridge Prison, one skeleton was subsequently shown to have injuries that were consistent with the historical records of Ned Kelly's wounds. How was that used in the identification of his remains?

Long after you are dead your skeletal remains can tell many stories about you. They can tell your sex, approximate age and physical traits indicative of your geographical region of origin, known as ancestry. An analysis of skeletal remains can tell what injuries you might have sustained throughout your life and they can also tell some of the diseases you might have suffered. In the hands of a skilled forensic anthropologist your bones can reveal things that might have been unknown to close family and friends. And, importantly, a forensic anthropologist can assist in the identification process.

So when the Pentridge Prison cemetery was investigated by consultant archaeologists TerraCulture and they found only one body buried there, a Victorian Institute of Forensic Medicine anthropologist with expertise in exhumations was requested by Heritage Victoria to assist in the recovery process. The body was later confirmed to be that of Ronald Ryan (see p. 116).

The Institute's involvement didn't stop there: when the three burial pits were later uncovered, between 2008 and 2009 (see Chapter 4), the remains of at least 34 individuals had to be recovered and examined. As the coffins with human remains were recovered from Pentridge they were transferred to the Institute.

The first stage of the Institute's work involved staff assigning an individual case number to each box and/or coffin of recovered human remains. The remains were then cleaned using warm water and soft brushes, before being analysed by standard anthropological techniques including description and measurement of different parts of the body. At the beginning of the work on the skeletal remains from Pentridge, the role of the anthropologist was to 'catalogue' the human remains (as opposed to identifying individuals).

All skeletal remains and associated evidence were photographed and CT-scanned as part of the documentation process. Surviving teeth and dentures were then independently examined by a forensic odontologist or dentist.

Estimating the numbers of individuals buried

Pit A (discovered in 2008) was found to contain three layers of coffins. These represented the reburials from the Old Melbourne Gaol of the remains of prisoners buried there which were transferred to Pentridge in 1929. There were an estimated 12 individuals in these coffins.

- Layer 1: Coffins 1–5 (minimum five individuals);
- Layer 2: Coffins 6–9 (minimum four individuals);
- Layer 3: Coffins 10–12 (minimum three individuals).

One case from Pit A showed some evidence of commingling, or skeletal remains from different individuals mixed together. The remains of five individuals were recovered in 2008 from Pit B, where no evidence of commingling was found.

As outlined in Chapter 4, Pit C was not uncovered until 2009, the following year. Archaeologists found 17 boxes as well as seven coffins, that contained human remains. The contents of each box or coffin were analysed as separate cases, for the purpose of examination. In most instances incomplete individuals were found but it could not be assumed that each box or coffin contained a single individual as, of the 24 boxes or coffins examined, eight showed evidence of commingling

In order to estimate the minimum number of individuals present in the collection, the anthropologist had to determine the bone that most frequently occurred. In this case it was the left collar bone (clavicle). By counting the number of these, it could be concluded that there were a minimum number of 19 individuals (including men and a woman) buried in Pit C. Looking at all three pits together, a minimum number of 34 individuals was determined to be buried, again based on the left clavicle.

The number of named individuals listed on a rough map of the burials that had been discovered, corresponded with the number of cases recovered from Pits A (12 cases) and B (five cases). However, Pit C was assigned 24 case files (there being 24 boxes or coffins containing skeletal remains), even though only 15 individuals were listed as being buried in that area on the rough map.

Skeletal analysis

Examples of cases that were relatively complete and had intact skulls were found in all the burial pits. Some cases, however, had only fragmentary skeletal remains with no skulls. Pit C, for example, had only one case that had a complete skull, but had nine cases that had partial skulls. Many of the bones were noted to be dark in colour, and some had different substances covering them. These included a clay-based soil and a lime-like substance.

Skeletal remains from Pit A had some pitch stuck to them. Only one case from Pit A had any surviving hair. The bones from one case were distinguished by being much lighter in colour and because of a strong naphthalene (mothball) odour. This may have been from the original embalming process.

Some bones from each of the three pits had shown evidence of incised or 'saw' marks which indicated that the individuals had been subjected to an autopsy or at least some form of internal post mortem examination (Fig. 5.1). Three cases (two from Pit A and one from Pit C) had non-human skeletal remains associated with them, although no attempt was made to determine what animals they were. In most cases, relatively poor preservation of the human skeletal remains limited the amount of detailed biological profile that could be developed. Where a determination could be made, the majority of individuals were Caucasoid/Caucasian adult males, although Pit A had evidence of an adult Asian male. And

Developing a biological profile

Part of the role of the forensic anthropologist is to develop a 'biological profile' which includes information about the person's ancestry, sex, age and stature (height) when they were alive.

- *Estimating the ancestry of the individual*
  An estimation of ancestry may be derived by assessing the presence or absence, or degree of development of skeletal and dental features (traits). An assessment of ancestry is typically undertaken only on adult remains because skeletal development is still occurring in children, and therefore it is usually very difficult to determine their ancestry. The region of the skeleton which most strongly reflects ancestral traits is the skull. There are also several head measurements that may be used to differentiate between populations. Depending on the preservation of the remains, a series of cranial measurements are taken and entered into a computer package which provides an ancestry assessment with related likelihood percentages.
- *Estimating the sex of the individual*
  Sexual differences of the human skeleton do not develop until after puberty. Therefore estimating the sex of juvenile skeletal material based on their shape alone is unreliable. There are two approaches to estimating the sex of an adult from the skeleton. These involve the assessment of shape (morphological) features of the skeleton and/or the use of measurements (metric criteria) in the form of indices (ratios of measurements), as well as 'discriminant functions' (often involving multiple statistics).
- *Estimating the age of the individual*
  From the foetal stage to around 25 years of age the human skeleton and dental areas undergo a continual process of development. Once a person has reached adulthood, the growing process ceases and the skeleton starts to degenerate. Therefore, examination of the skeletal and/or dental remains provides a more accurate age for younger as opposed to older individuals. In most cases age categories are presented as a range (Table 5.1).

Table 5.1: Age ranges

Juvenile	
Foetus	9 weeks to before birth
Infant	0– <2 years
Child	2–12 years
Adolescent	13– <20 years
Adult	
Young adult	20–35 years
Middle adult	35–50 years
Old adult	50+ years

- *Estimating the stature of the individual*
  Stature is estimated using maximum length measurements of long bones such as the thigh (femur), lower leg (tibia), arm (humerus) and forearm (radius and ulna). Measurements are taken using an instrument called an osteometric measuring board. Depending on the skeletal element and the ancestry of the remains, a specific mathematical formula (a regression equation) is employed to determine approximate stature.

Soren Blau

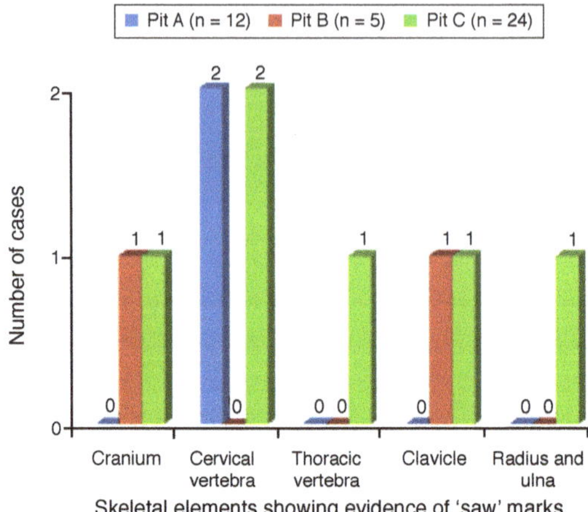

Fig. 5.1: Number of cases from Pits A, B and C at Pentridge Prison showing evidence of 'saw' marks. Victorian Institute of Forensic Medicine.

while most individuals were either young or middle-aged, Pit C had evidence of a young teenager and a couple of females (Figs 5.2, 5.3, 5.4). The age, sex and general place of origin of ancestry of a skeleton can be determined by several factors (see 'Developing a biological profile', p. 53).

No attempts were made to estimate the height of individuals, although measurements of relevant long bones were taken, as height was not seen as useful data for identification purposes. Many cases had intact dental remains and one individual from Pit B had both upper and lower dentures, but only one individual showed signs of having had dental restoration done during his life. Several bones showing evidence of disease and traumatic injuries were identified among the remains from all pits from Pentridge. These included:

Pit A:

- healed fracture to the lower left leg;
- cases of possible healed rib fractures.

Pit B:

- healed fracture to the lower left leg.

Pit C

- shotgun wound in the tibia (shin bone);
- healed fracture on the ribs;
- the formation of new bone on the surface of the bone, known as exostosis (on the upper arm bone or humerus, feet and pelvis);
- possible infection of the bone (osteomyelitis, distal humerus);
- degeneration of the bone into a hard ivory-like mass – this occurs when the cartilage at the joint is worn down (eburnation, on the forearm bone, i.e. radius);
- development of bone spurs along the edge of the joints (osteophytes, on the clavicle, vertebrae and sacrum);

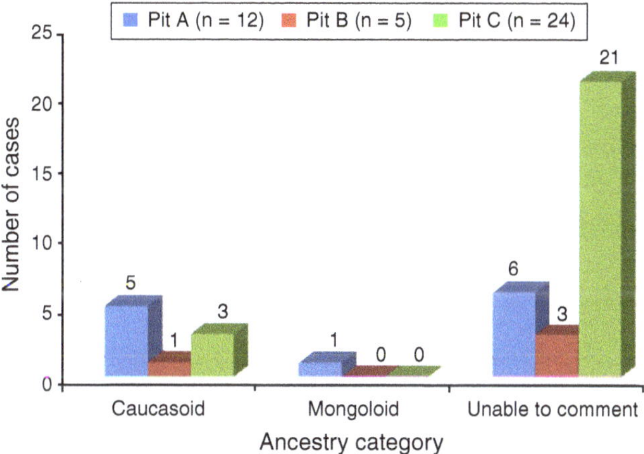

Fig. 5.2: Estimation of ancestry, Pits A, B and C, Pentridge Prison. Victorian Institute of Forensic Medicine.

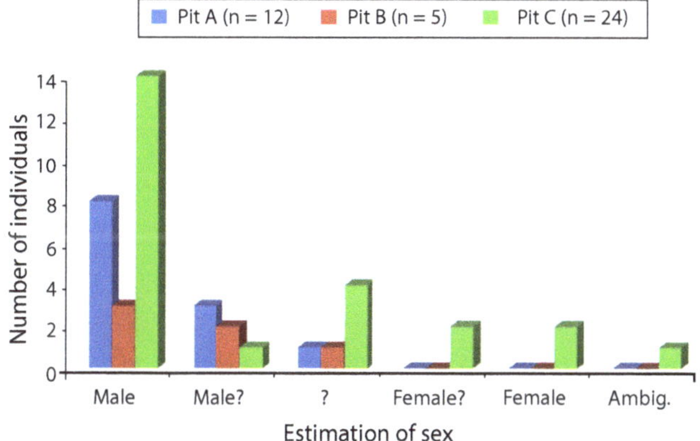

Fig. 5.3: Estimation of sex, Pits A, B and C, Pentridge Prison. Victorian Institute of Forensic Medicine.

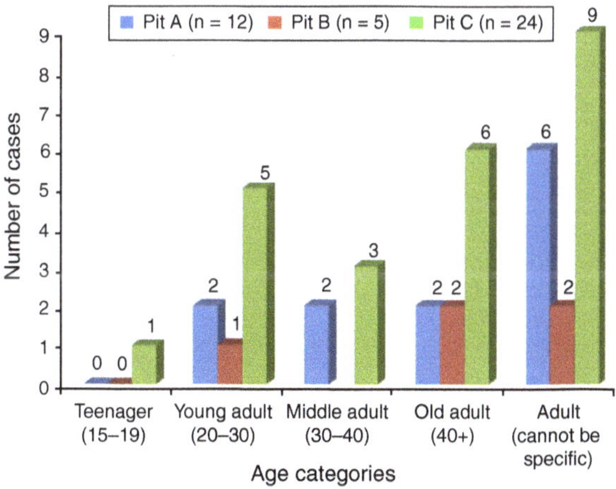

Fig. 5.4: Estimation of age at death, Pits A, B and C, Pentridge Prison. Victorian Institute of Forensic Medicine.

- a disorder of the joint in which a loose piece of bone and cartilage separates from the end of the bone because of a loss of blood supply (osteochondritic) lesions (on one of the bones of the feet, i.e. navicular).

In October 2010, the then Attorney-General Rob Hulls announced funding for the attempted identification of Ned Kelly from among these remains. Following the successful matching of mitochondrial DNA from bone samples with samples provided by a living relative of Kelly (see Chapter 8) anthropological analysis of the remains from the individual known as case 3081/10 was continued and found several features supporting the DNA conclusion that the remains were of Ned Kelly. Of most significance are the following skeletal alterations:

- a defect on the left elbow-end of the arm bone (humerus) indicating the bone was healing following an injury (Fig. 5.5);
- 'saw' marks on the mid-shafts of the left forearm bones (radius and ulna). The proximal parts of the radius and ulna were not present in the collections (i.e. were absent post mortem) (see Fig. 5.6);
- a circular defect with a diameter of about 8 mm located on the lateral aspect of the proximal right shin bone (tibia). This defect was indicative of a gunshot wound (see Fig. 5.7);
- a defect including a fracture on the proximal joint surface of the right big toe (see Fig. 5.8);
- a piece of the occipital bone from the skull (see Fig. 7.2).

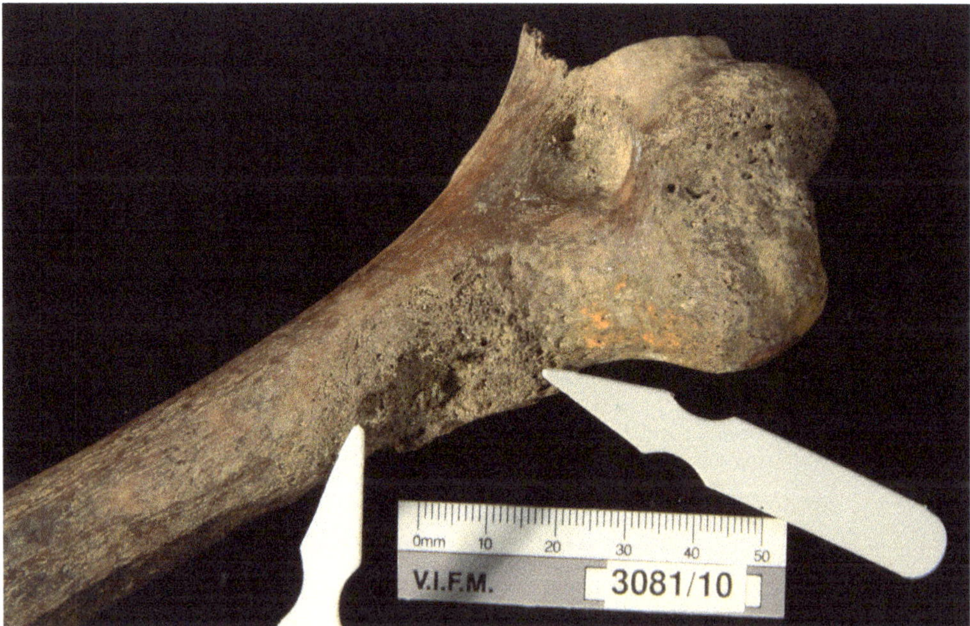

Fig. 5.5: View of the part of the left upper arm bone (humerus) showing a defect with indications that the bone was healing following an injury. Victorian Institute of Forensic Medicine.

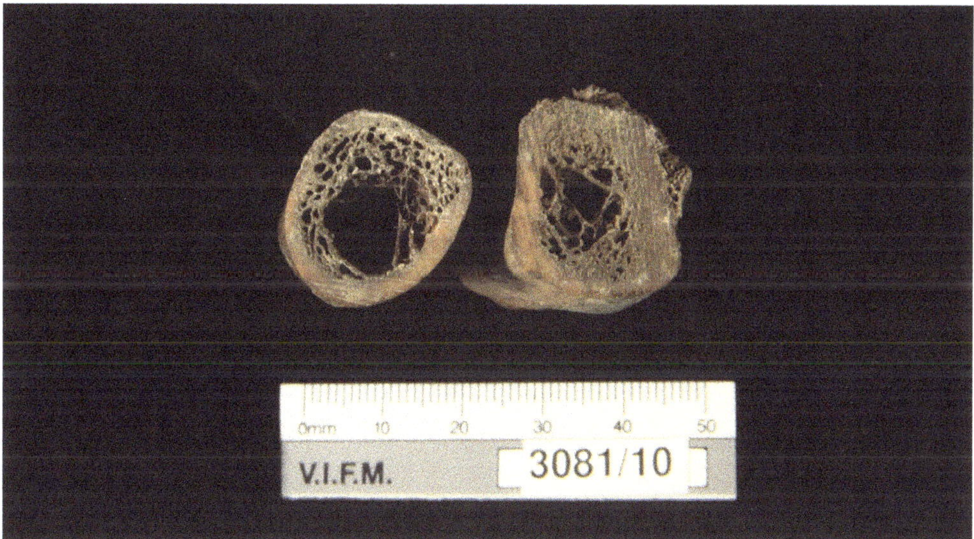

Fig. 5.6: View of the mid shaft sections of the left forearm bones (ulna and radius) showing evidence of having been sawn. Victorian Institute of Forensic Medicine.

The pathological and traumatic changes to the bones described above were compared with the historical descriptions of Kelly's injuries following the shooting at Glenrowan. Sergeant Arthur Steele of the Wangaratta Police Station is attributed with having injured Kelly by shooting at his unprotected legs during the siege at Glenrowan, which ultimately

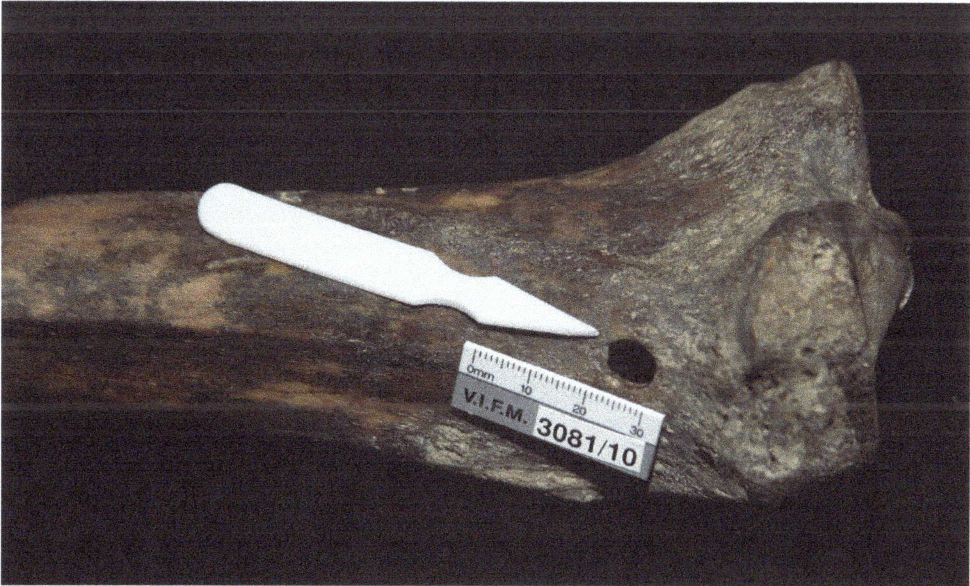

Fig. 5.7: View of the lateral aspect of the proximal right shin bone (tibia) showing an approximately 8 mm diameter circular defect with smooth edges, indicative of a healing shotgun pellet wound. Victorian Institute of Forensic Medicine.

Fig. 5.8: Articular surface of the right big toe showing evidence of a fracture. Victorian Institute of Forensic Medicine.

resulted in Kelly's capture and arrest. In a statement dated one month after the Glenrowan siege, Sgt Steele described the event:

> I ran towards him to within 15 yards … I immediately fired at him on the outside of his right leg, he staggered and his hand dropped, he again tried to raise the revolver Where I again fired at him about 10 yards distant, on the hand and thigh which were in a line (Steele 1880).

The medical officer who examined Ned Kelly shortly afterwards, Dr A. Shields, noted:

> The principal injuries are: Firstly, a severe bullet wound near the left elbow; there are two openings, one above the other, below the joint, the two apertures having probably been caused by the bullet traversing the arm when bent. Secondly, the right hand has been injured near the root of the thumb. From this I removed a large slug-shot. Thirdly, on the right thigh and leg there are several wounds, caused by the same kind of shot … Fourthly, the right foot has received a severe injury. The track of the ball here is marked by two openings, one on the top of the ball of the great toe, and the other on the sole of the foot, and the bone is damaged (Shields 1880).

Similar injuries to Ned Kelly's

The alteration observed near the left elbow of 3081/10 did not provide actual evidence of a firearm wound, but it was possible that the changes to the skeleton observed there were the result of the body reacting to such an injury. The fact that the joint of the left elbow was spared is supported by Shield's description of the injury to the left elbow. Also, the fact that the bones of the upper left forearm (proximal radius and ulna) were missing, and were most likely removed by sawing, suggests that the injured arm was dissected in some form of post mortem examination.

The circular defect on the right lower leg also showed a feature of a shotgun pellet wound and a CT scan of the right shin bone (tibia) showed evidence of pellets and metallic fragments associated with this injury (see Fig. 5.9). Two shotgun pellets were subsequently recovered from the lower tibia (and they must have both entered through the one hole). A CT scan also showed evidence of metallic fragments in the right big toe (metatarsal) (see Fig. 5.10), which also fitted with Dr Shield's description of Kelly having been shot in the right foot.

The combination of the mtDNA and anthropological findings allowed the incontrovertible conclusion that the skeletal remains were those of Edward 'Ned' Kelly.

Prior to the advent of DNA analysis, a court faced with i) the general anthropological findings (sex, age), ii) the specific anthropological findings of the injuries, iii) the historical information about Ned Kelly's injuries, and iv) the circumstances of the discovery of the

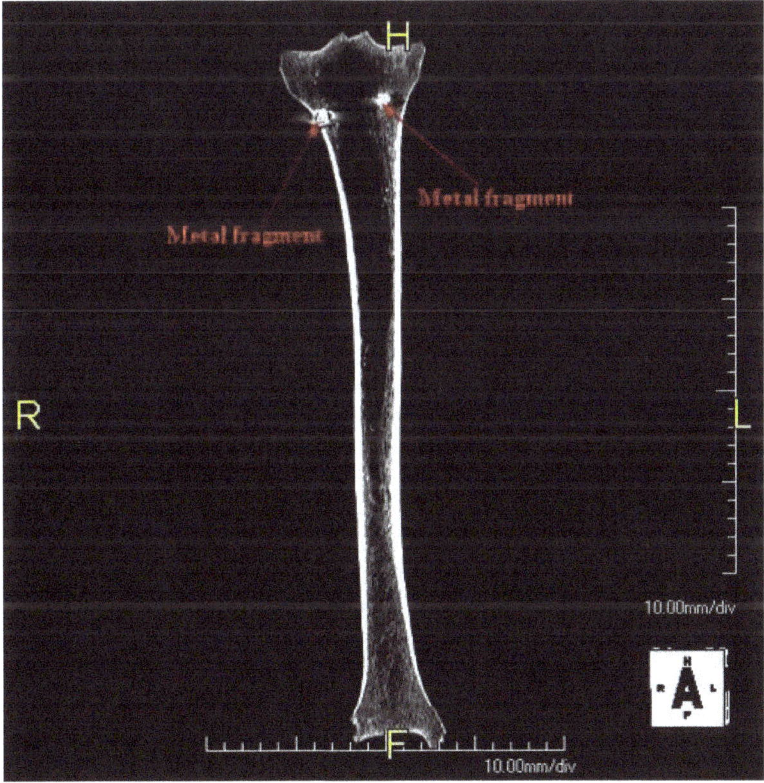

Fig. 5.9: CT scan of the right lower leg (tibia) showing shotgun pellets and metallic fragments associated with the injury. Victorian Institute of Forensic Medicine.

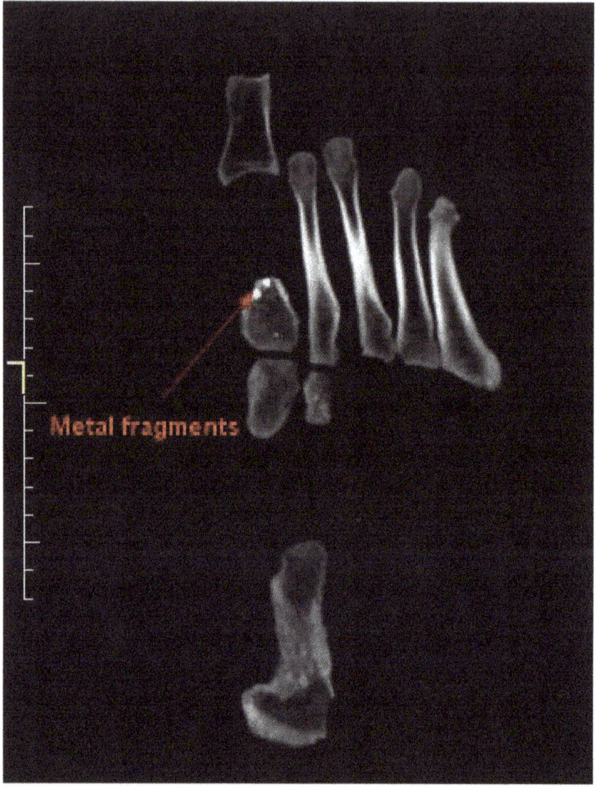

Fig. 5.10: CT image of the bones of the right foot showing metallic fragments in the partial big toe (metatarsal). Victorian Institute of Forensic Medicine.

remains (i.e. from Pentridge and the information of the rough map) would, in our view, have concluded that these remains were those of Edward 'Ned' Kelly.

Note: the majority of this chapter is based upon S. Blau *et al.* (2014) The contributions of anthropology and mitochondrial DNA analysis to the identification of the human skeletal remains of the Australian outlaw Edward 'Ned' Kelly. *Forensic Science International* **240**, e11–e21.

References

Shields A (1880) The official medical report. http://www.kellygang.asn.au/documents/N80/80_06_30_Argus2.html (accessed 08/01/13).

Smith J (2011) Losing the plot: archaeological investigations of prisoner burials at the Old Melbourne Gaol and Pentridge Prison. *Provenance: The Journal of Public Record Office Victoria* 10. http://prov.vic.gov.au/losing-the-plot (accessed 03/09/13).

Steele LM (1880) Statement re Glenrowan outrage. VPRS 4966 Consignment P0 Unit 1 Item 2, Record 15 Document: Sergeant Steele re: Glenrowan Outrage, 20/07/1880. http://prov.vic.gov.au/whats-on/exhibitions/ned-kelly/the-kelly-story/brought-to-justice/queen-v-edward-kelly-murder-file (accessed 08/01/13).

Chapter 6
Analysis of the skull using odontology and craniofacial superimposition

Richard Bassed, Anthony Hill and Noel Woodford

A single tooth was used to prove that the skull, thought to be Ned Kelly's, being examined at the Victorian Institute of Forensic Medicine had come from the Old Melbourne Gaol graveyard. But when images of the skull were laid over photographs of Ned's face and death masks, as well as those of other executed prisoners, there were two possible matches.

A dry human skull weighs about 1 kg – approximately the same as a litre of milk. When holding such a poignant object in your hand one of the first things you notice is the myriad of irregular 'suture' lines crossing the dome of the skull and extending onto the base of the skull; these represent the lines of fusion of many of its 28 separate bones. These separate bones slowly join together as a person grows and ages, beginning during the third month in utero, and ending with final fusion of a bone at the base of the skull at 16–17 years of age.

The skull, perhaps more than any other part of the body, is thought to represent the person who once breathed and spoke, and whose eyes looked out of the now empty sockets. As we gaze at this reminder of our own mortality, we wonder about the individual whose brain once occupied the empty vault, what they thought and did. Were they good or bad? Vengeful or forgiving? And perhaps first of all, we wonder what they might actually have looked like.

While a skull can give us an idea of the shape of a person's head and face when they were alive, this is by no means an exact science and there is scope for artistic interpretation. Despite what is shown on many 'CSI' type television programs, we have no way of proving definitively from the skull alone what the size and shape of the nose was, for example, or the ears, or the colour of the eyes. We certainly have no idea how fat or thin a person was or what colour hair they had.

What forensic science is able to do, however, in certain situations, is to either include or exclude the possibility that the skull belongs to a specific person. If it can be shown that a skull is either too large or too small for a particular head, then that person can be excluded as a possible match. And this was one of the techniques used by the forensic anthropologist and odontologists at the Institute when they undertook craniofacial superimposition of the skull thought to be Ned Kelly's, comparing it to death masks and photographs of executed prisoners.

When forensic dentists, anthropologists and pathologists who work at the Institute were handed the skull with the name 'E. Kelly' inscribed on it that Tom Baxter had kept hidden for just over 30 years, the team had four key questions to try to answer. All the questions contributed to establishing the skull's identity, and thus required an investigation that attempted to place the skull at certain critical points in time.

1 Was the 'Baxter skull' the same skull that was stolen from the Old Melbourne Gaol in 1978?
2 If yes, was the 'Baxter skull' the same as the 'Kelly' skull that had been held at the Institute of Anatomy from the 1930s until 1972?
3 If yes, was the 'Baxter skull' removed from the grave of an executed prisoner at the 1929 Old Melbourne Gaol Labour Yard exhumation?
4 If yes to all of those, was the 'Baxter skull' that of Edward 'Ned' Kelly?

To answer these questions, the team had to rely on a combination of scientific methodology and historical research. While the historical team was sifting through the enormous amount of archival data, the Institute's scientific team was employing various forensic examination methods to attempt to establish the authenticity of the skull's identity by establishing its 'provenance', as researchers call it.

The skull was first examined to determine such things as age, sex, ancestry and signs of injury or medical intervention (see Chapter 3). Old photographs recovered by the historians were compared with the skull and as many death masks as were available of executed prisoners from the Old Melbourne Gaol were also compared to it. Small samples of bone were taken from the skull in an attempt to extract DNA.

Making death masks

The method used for making death masks involved techniques familiar to many dentists and to anyone who has had a dental mould made for a sport mouthguard. The deceased, whose hair and beard were shaved, was placed upright in a chair with the head extending through a hole in a workbench, allowing access to the entire head and neck. Any areas of obvious difficulty, such as the nostril and ear openings, were blocked out with damp paper then the entire head was painted with a thin plaster and water mixture. A string was placed in this thin covering, extending from back to front of the head, across its centre.

A thicker plaster mix was applied over the thin plaster layer to create strength, and was allowed to partially set. The previously placed string was then drawn out through the stiffening plaster, thus allowing the mould to fall into two halves. Once these two halves had been separated they could be rejoined and the mould filled with plaster to create the bust. Kelly's death mask was made by Maximilian Kreitmayer, proprietor of a waxworks museum in Bourke St, Melbourne.

Often, many copies of executed prisoners' likenesses were constructed. Sometimes these plaster facsimiles were sold to the public outside the prison; for example, there are anecdotal reports that even the Melbourne Gaol Governor John Buckley Castieau may have run a business supplying copies of death masks to his acquaintances. Many of the death masks were examined by now-discredited phrenology experts – the 'lumps and bumps' interpreters – who would ascribe various personality traits to an individual on the basis of head shape.

Richard Bassed

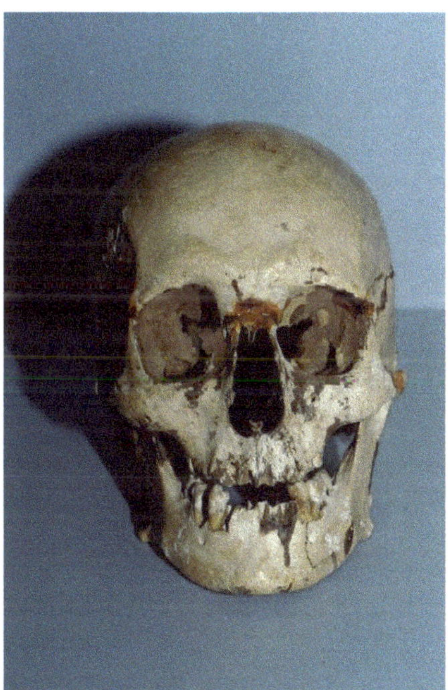

Fig. 6.1: Front view of the 'Baxter skull'. Victorian Institute of Forensic Medicine.

The skull: first impressions

The first thing noted about the skull presented to the Institute by Tom Baxter was its condition (see Fig. 6.1). It was given to us in a dilapidated make-up case and showed signs of being heavily weathered, as well as evidence of having been repaired with glue in the region of the left cheek bone, the junction of the nasal bones with the forehead and the upper right rim of the right eye socket. The bone was a pale brown colour, quite dry and light, indicating that considerable time has passed since death (fresh skulls – i.e. those of individuals who have died recently – are heavy and somewhat greasy to the touch due to a high organic material content, which slowly disappears over time). On the left side of the skull, in the region of the temple, 'E. Kelly' had been penned in black ink.

Forensic examination

The anthropologists' initial examination (see Chapter 5) noted that the skull appeared to be that of an adult Caucasoid male and that, while largely intact, several bones of the skull were missing, including:

- the sphenoid bone (behind and below the nose and forming part of the base of the skull);
- the ethmoid bone (behind the roof of the nose and containing the nerves that provide the sense of smell);
- the central part of the right zygomatic arch (cheek bone).

All of this damage was thought to have occurred after death, probably as a result of careless handling of the skull over a long period. There was no evidence of any autopsy procedure on the skull.

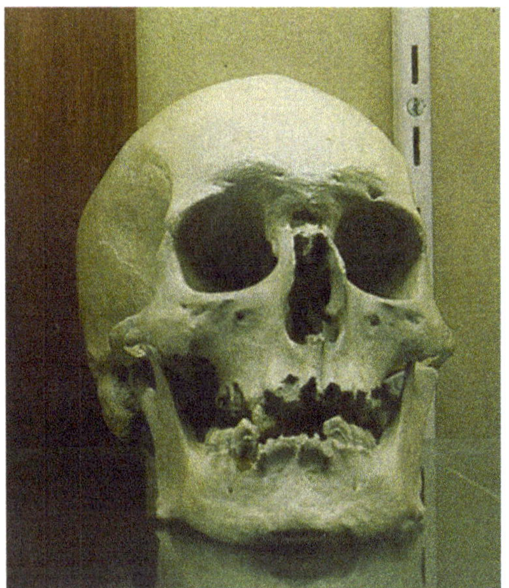
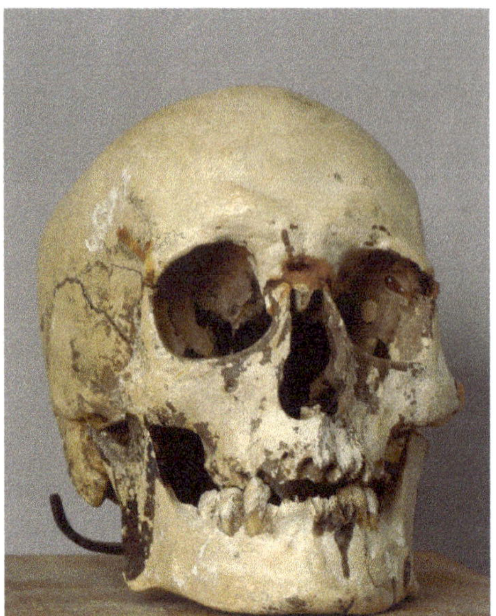

Fig. 6.2: Front view of the skull from the Old Melbourne Gaol postcard (on left), compared to the 'Baxter skull' (right). Victorian Institute of Forensic Medicine.

Forensic dentists (odontologists) examined the skull and reported that all of the teeth from the upper jaw were missing. This had happened after death, as there was no evidence of any healing in the tooth sockets as would have normally occurred if any teeth had been lost while the person was still alive. The lower jawbone (mandible), however, still had eight teeth present. The jaw also articulated (i.e. moved easily) within its joint, consistent with the skull and jaw being from the one person.

Question 1. Was the 'Baxter skull' the same skull that was stolen from the Old Melbourne Gaol in 1978?

Addressing the first question, and in order to determine if the 'Baxter skull' was indeed the one that had been on display in the Old Melbourne Gaol, the team looked for and eventually found a postcard from the mid 1970s which showed the skull in a glass case alongside Kelly's death mask. When this image was laid alongside the 'Baxter skull' (see Fig. 6.2) the similarities and differences could be visually examined closely.

The similarities between the two images were:

- the shape of the jaw, particularly with reference to the flaring of its angles;
- the shapes and number of teeth present in the jaw;
- the shapes of the tooth sockets in both upper and lower jaws where teeth had been lost after death;
- the shape of the nasal cavity;
- the brow ridges (supra-orbital) and location of the passages for nerves of the face (infra-orbital and supra-orbital foramina).

And the differences were:

- the skull had been fractured and repaired with glue in several areas. These repaired areas were not visible in the Old Melbourne Gaol postcard;
- the skull showed no evidence of any upper teeth, yet there was a single upper tooth visible in the postcard;
- the skull displayed staining and weathering not visible in the postcard.

Despite the noted differences, it was concluded that they were due to weathering over an extended time and to poor storage, and that the postcard image and skull in the Institute's possession were indeed one and the same.

Question 2. Was the 'Baxter skull' the same as the 'Kelly' skull that had been held at the Institute of Anatomy from the 1930s until 1972?

Efforts to establish the provenance of the 'Baxter skull' back to the 1930s involved comparing the skull with a plaster replica made by the National Museum of Australia in the early 1930s, when the skull was reportedly sent there from the Institute of Anatomy. The replica was photographed in numerous anatomical positions and a visual comparison was then made with the 'Baxter skull'.

There were several striking points of similarity between the National Museum replica and the 'Baxter skull' (see Figs 6.3, 6.4, 6.5):

- overall size and shape of the cranium was identical in both the 'Baxter skull' and the replica;

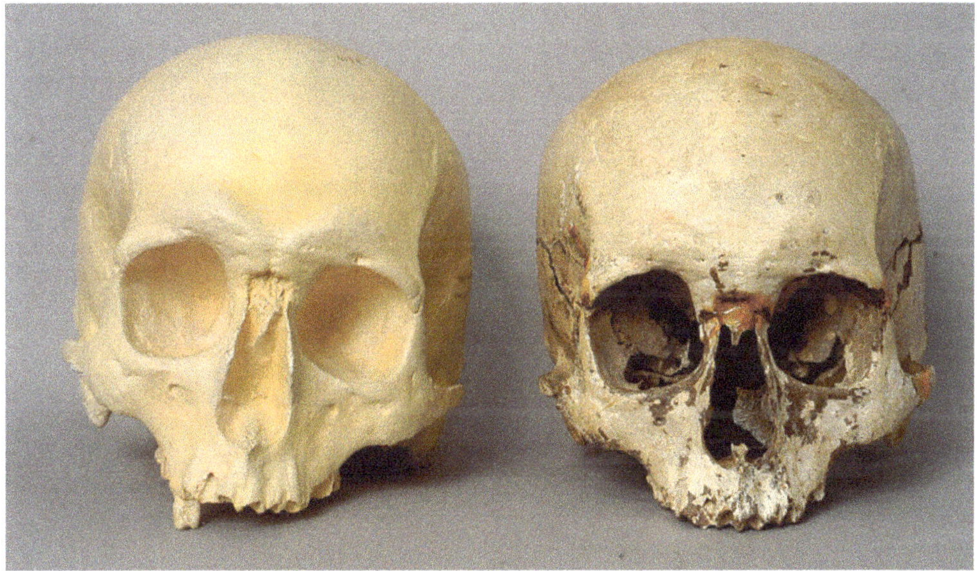

Fig. 6.3: Front view of the skull replica from the National Museum of Australia (on left), compared to the 'Baxter skull' (right) – frontal view. Victorian Institute of Forensic Medicine.

- upper jaw structure and tooth socket shape;
- shape of nasal cavity and position of nasal spine;
- shape of the passages for nerves of the face;
- position and size of the passages for nerves of the face (infra-orbital and supra-orbital foramina);
- position and shape of the mastoid processes bones behind the ears;
- shape and nature of the joins of the skull;
- size, shape and nature of the roof of the mouth;
- morphology and number of tooth sockets in the jaw.

And the differences:

- post mortem loss of bony structure from the right side of the base of the 'Baxter skull', including the majority of the sphenoid bone (an open wing-shaped bone located behind the eye sockets that extends to form the base of the skull);
- the right cheek bone (zygomatic arch) of the 'Baxter skull' was partially missing, while it was intact in the replica;
- the upper right premolar tooth was present in the replica but missing in the 'Baxter skull';
- severe weathering and repair of damage using glue in the 'Baxter skull' was not evident on the replica.

While the visual comparison was able to determine a high number of consistent points of similarity and no incompatibilities which could not be explained, the defining comparison involved the base of the roof of the mouth. This showed an identical number, shape, size and position of tooth sockets between the 'Baxter skull' and the replica. From this comparison the Institute scientists were able to conclude that the 'Baxter skull' was the same as that handed to the National Museum of Australia in the early 1930s and was the skull from which the plaster replica was made.

Question 3. Was the 'Baxter skull' the same one removed from the grave of an executed prisoner at the 1929 Old Melbourne Gaol Labour Yard exhumation?

If it could be shown that the 'Baxter skull' had been exhumed from the Old Melbourne Gaol Labour Yard in 1929, then it would be established that the skull did indeed belong to an executed prisoner. The establishment of this as fact would mean that the number of possible original owners of the skull was significantly reduced, and although this would not prove that the skull was that of Ned Kelly, it would make the task of final definitive identification a much simpler task.

The historical team fortuitously discovered a newspaper report covering the exhumation of prisoner remains from the Old Melbourne Gaol Labour Yard in 1929, that led to the discovery of a photograph of an individual holding a skull which he said was that of Ned Kelly. Chris Ott brought in the photograph and a sworn statement from Alex Talbot in which he said he had in his possession a tooth taken from the skull (see Fig. 2.4).

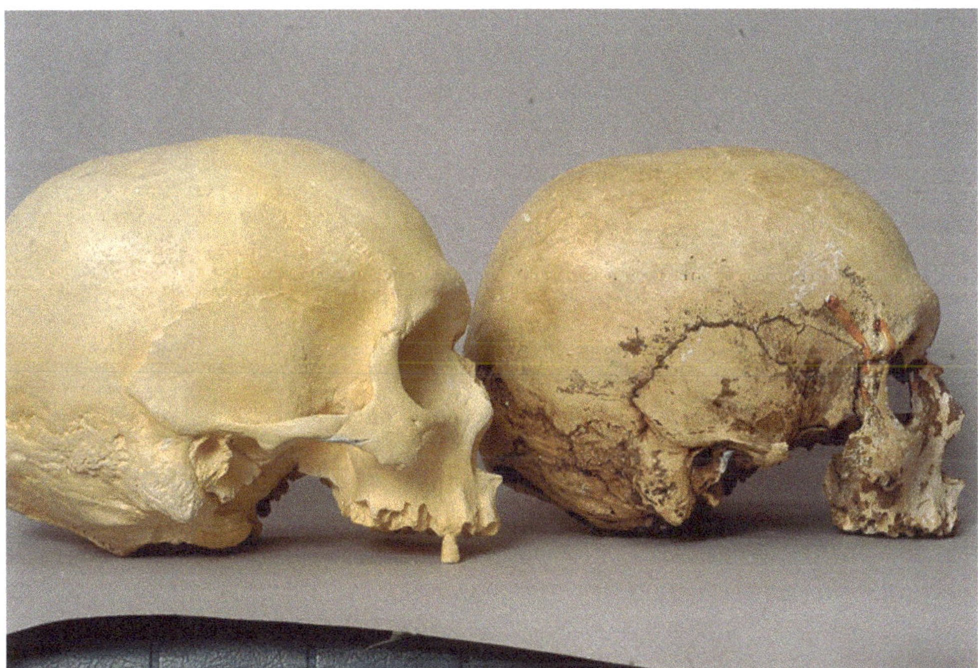

Fig. 6.4: Right side view of the skull replica from the National Museum of Australia (on left), compared to the 'Baxter skull' (right) – side view showing missing central portion of the zygomatic arch and missing upper right premolar tooth. Victorian Institute of Forensic Medicine.

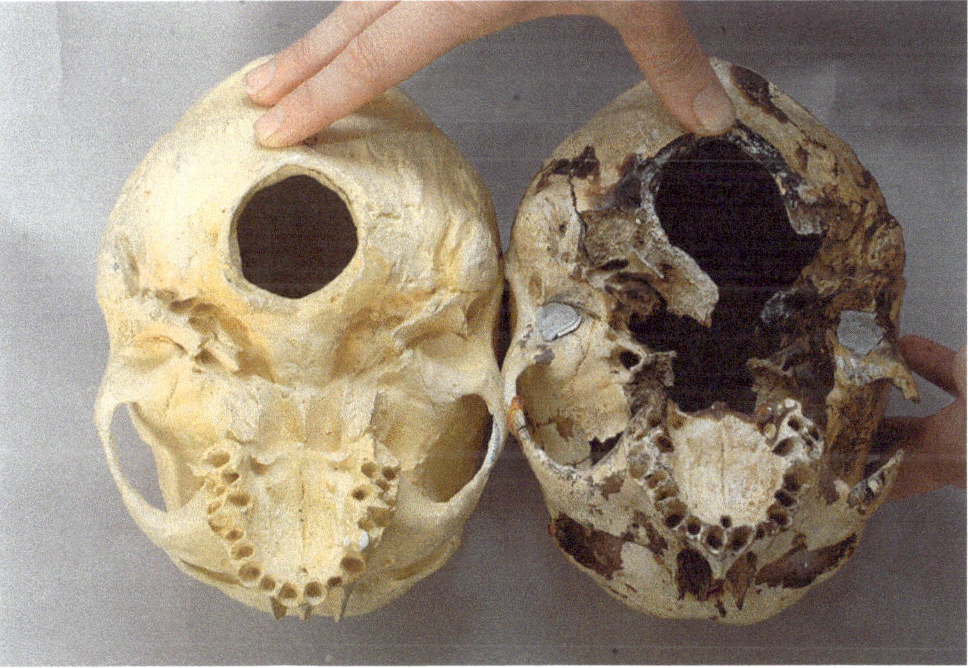

Fig. 6.5: View from below of the skull replica from the National Museum of Australia (on left), compared to the 'Baxter skull' (right) – roof of the mouth. Victorian Institute of Forensic Medicine.

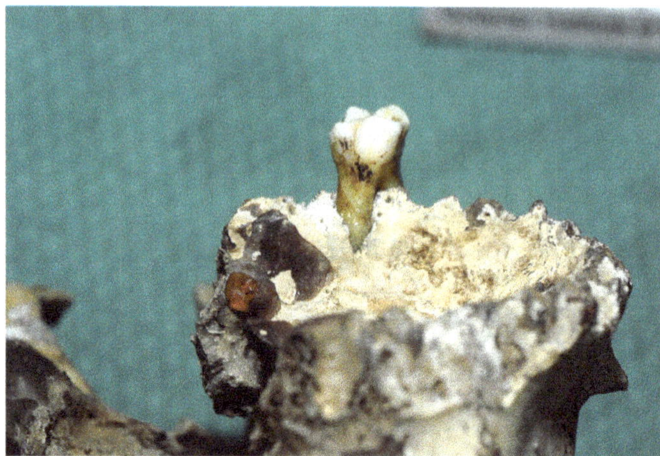

Fig. 6.6: The recovered tooth within the bony socket of the upper jaw of the of 'Baxter skull'. Victorian Institute of Forensic Medicine.

Much to the odontologists' delight and surprise, the missing tooth was discovered to still be in the care of the family of the person who originally took it. The tooth was an upper right first molar with two roots. These teeth normally have three roots but one had been fractured off, perhaps at the time it was forcibly removed from the skull during the 1929 exhumation. The roots of the teeth and the bony tissue surrounding them are genetically determined, so the roots are formed in unique shapes for every person and the bony sockets into which the roots fit will only admit that tooth which is genetically determined to be there. This is especially true of teeth with multiple roots of differing shapes and angulations.

Amazingly, like the glass slipper fitting Cinderella's foot, the tooth fitted perfectly into a socket in the upper jaw of the 'Baxter skull' (see Figs 6.6 and 6.7).

The newspaper photograph of the individual holding the skull from the 1929 exhumation site was of quality insufficient to allow scientific comparison using superimposition

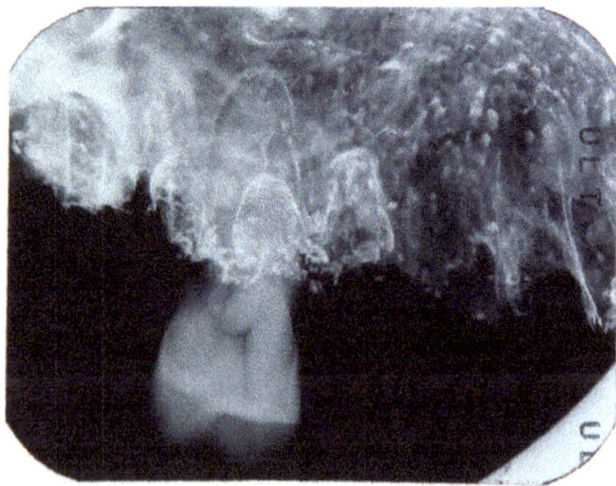

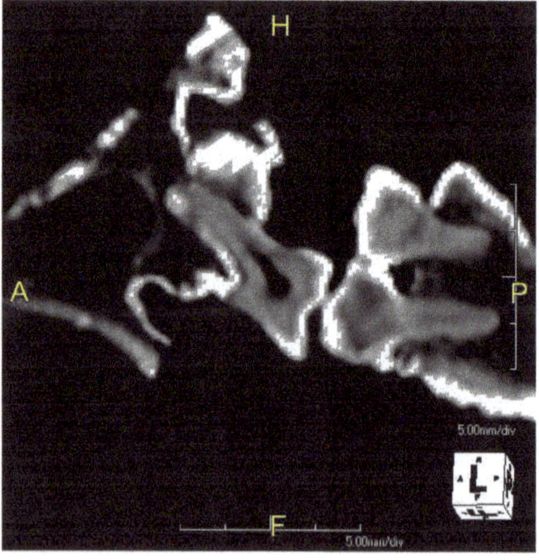

Fig. 6.7: Radiographic image and CT image of tooth within the 'Baxter skull'. Victorian Institute of Forensic Medicine.

techniques, however. But there were distinct similarities between the shape of the skull in the photograph and the 'Baxter skull', and no dissimilarities were identified.

The fact that the tooth fitted perfectly into the socket in the 'Baxter skull' confirmed that it was indeed the remains of an executed prisoner buried in the Labour Yard. This reduced the number of possible individuals whose skull it could have been to those buried within a specific area of the Labour Yard. But it did not yet prove this was the skull of Ned Kelly.

Question 4. Was the 'Baxter skull' that of Ned Kelly?

Having established that the 'Baxter skull' belonged to an executed prisoner who had been buried in, and subsequently exhumed from, the Old Melbourne Gaol Labour Yard, the next task was to try to definitively place that skull in the head of one individual.

The available information consisted of the plaster death masks, or busts, made of 22 prisoners shortly after their execution and the photographs of three more taken during their lives. The data were incomplete: for several prisoners death masks had either not been made or had gone missing, and the available photographs were not the best quality.

Detailed photographs were taken of the 'Baxter skull' and of all the death masks, using standard forensic digital photographic techniques with an accurate scale in place. It was vitally important to ensure that identical distances, camera angles, lenses and scaling was achieved in each image so that the comparisons were made in a reproducible way. The images were then imported into a computer program.

Superimposition appears much more exciting on popular TV crime shows, where advanced computer modelling spins a skull in all directions on a large transparent screen in the middle of a meeting room. But in reality the Institute researchers use a regular off-the-shelf product – Adobe Photoshop®. The scales in each image were carefully measured and sized so that each image was equivalent in dimensions, then both side and frontal views of skull and death mask were superimposed to look for points of similarity or difference.

Particular features that we examined and compared included the following:

- whether or not the skull fitted inside the dimensions of the death mask;
- the position of the eye sockets of the skull in relation to the eyes of the death mask, and the relationship to the nasal cavity and nose and cheek bones;
- the jaw line, chin position, and position of the teeth;
- the relationship between the brow ridges and shape of the forehead;
- an overall impression of fit.

Further comparisons were made using much more sophisticated three-dimensional computed tomography (CT-scanning) superimposition software (Toshiba Fusion 7D®). This rather expensive and time-consuming method was reserved for the death masks that were considered possible matches. It involved superimposing three-dimensional CT images of the skull and death mask – essentially putting a 3D model of the 'Baxter skull' inside the 3D model of a death mask on a computer screen. This was done to see if further eliminations could be made.

Two matches! Kelly and Deeming

Only two of the 22 death masks or busts were assessed as being possible fits for the skull. All others were excluded on the basis that the skull was too large, too small or anatomical landmarks failed to align in an appropriate way (see Fig. 6.10). The two death masks that showed a reasonable degree of fit were those of Ned Kelly and another executed prisoner who, quite amazingly, had reportedly been buried right next to him in the Old Melbourne Gaol Labour Yard, Frederick Deeming (see Figs 6.8 and 6.9) (see p. 117).

This was frustratingly close. Superimposition of CT images of the death masks with the skull in 3D supported these conclusions, but we were unable to discriminate between Deeming and Kelly in terms of which one was a better fit.

Comparison of the skull with the photographic portraits of a further three executed prisoners didn't provide any strong alternatives as the photographs had no measurable scale and could not be accurately resized to the correct dimensions. Notwithstanding, researchers attempted resizing the images based on similarities to anatomical landmarks such as the nose opening and the jaw. But none of the photographs was seen to be a reasonable fit (see Fig. 6.10).

And that was as far as the science of photographic and CT craniofacial superimposition was able to advance the investigation. Again, unlike TV forensic crime shows, there is no strong scientific evidence that this technique can successfully be used to identify an

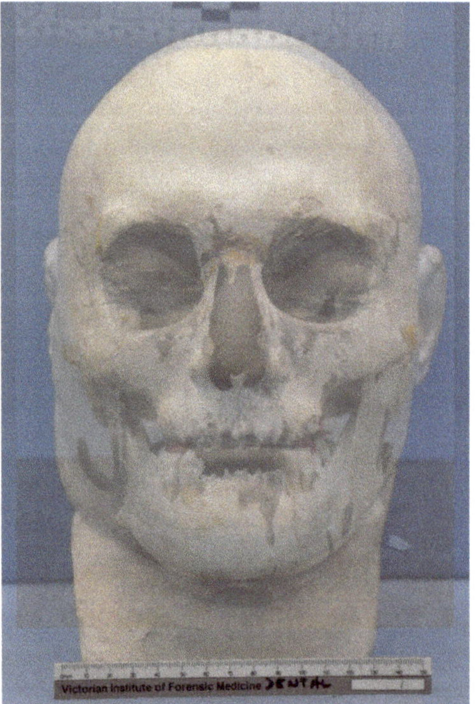

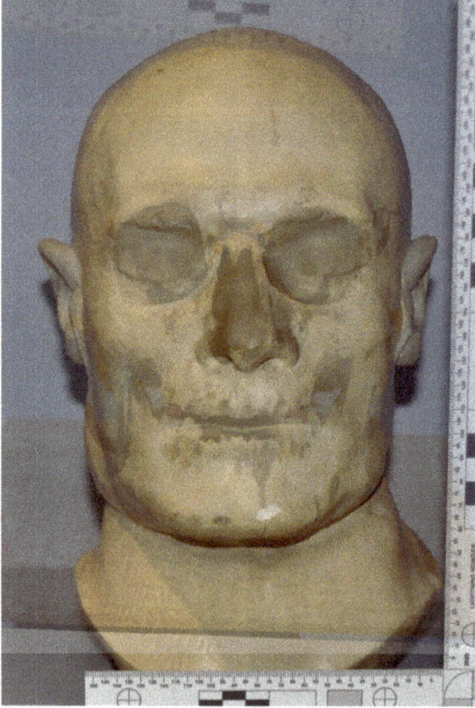

Fig. 6.8: Superimposition showed that Ned Kelly's death mask (left) was a close fit to the skull. Frederick Deeming's death mask (right) was also a close fit. Victorian Institute of Forensic Medicine.

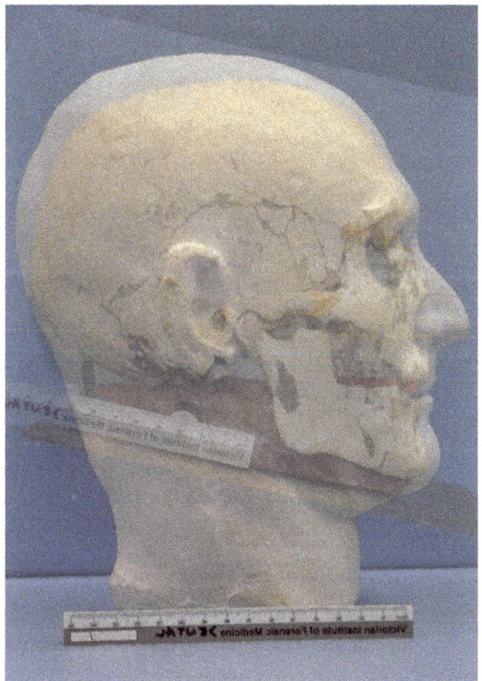
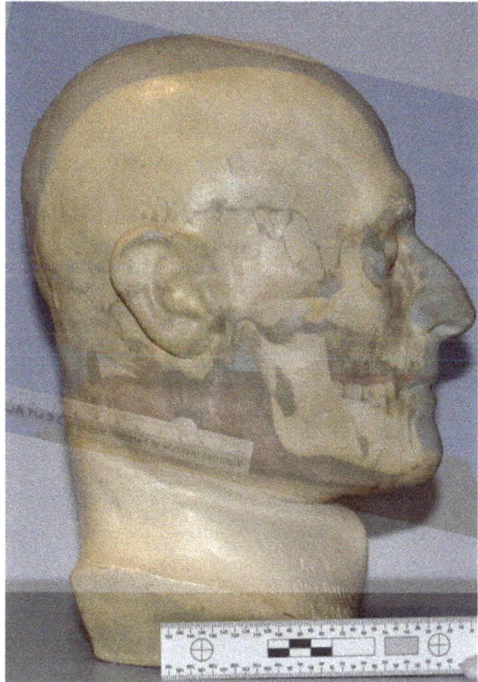

Fig. 6.9: Superimposition of CT images of the death masks of Kelly (left, reversed for comparison) and Deeming (right) could not determine which was a better fit for the skull. Victorian Institute of Forensic Medicine.

individual. Rather, it can be used to exclude individuals when seeking identification. To date there has been no reliable research showing that the technique alone can readily establish the identity of an unidentified human skull when compared with photographs of the head of known individuals before death, let alone with their death masks.

There is also no research showing that each human skull has such a unique shape that it can be used to positively identify an individual in the way fingerprints can be used, despite what is sometimes claimed in cases of identifying dead bodies by comparing skulls to photographs. So even if it had been possible to identify one match between skull and death mask or photo, it would not be acceptable to declare a positive identification, except in very rare cases where highly unusual anatomical features might be present.

It was actually quite a rare thing for an odontological team to be comparing a skull to death masks, and being able to do so in ideal situations with alignment of images. It was also different from the standard work of seeking to identify an individual by a skull in that the possible number of matches was limited to those buried at the Old Melbourne Gaol – in normal forensic casework the number of possible identities could be quite large.

While the study only managed to narrow the identity of the skull down to two possible death mask matches, with the proviso that there were several executed prisoners buried in the Labour Yard for whom no death masks had been made, it was useful in that it demonstrated that the possibility of the skull belonging to Ned Kelly could not be discounted.

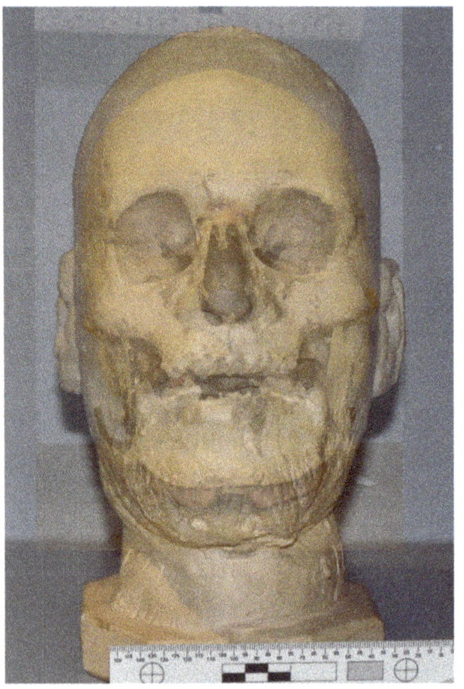

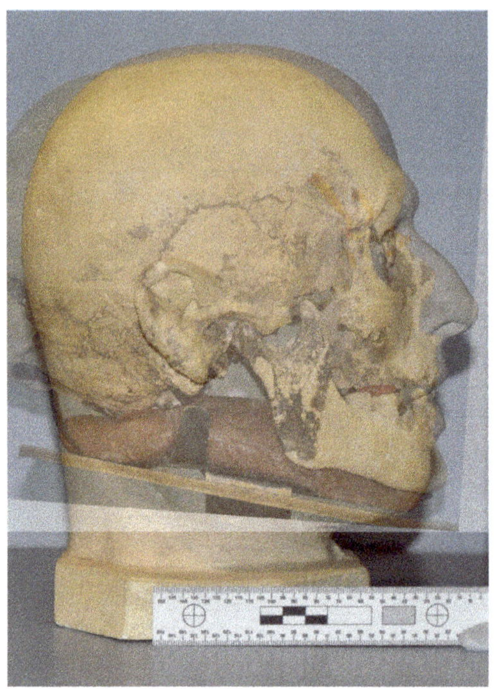

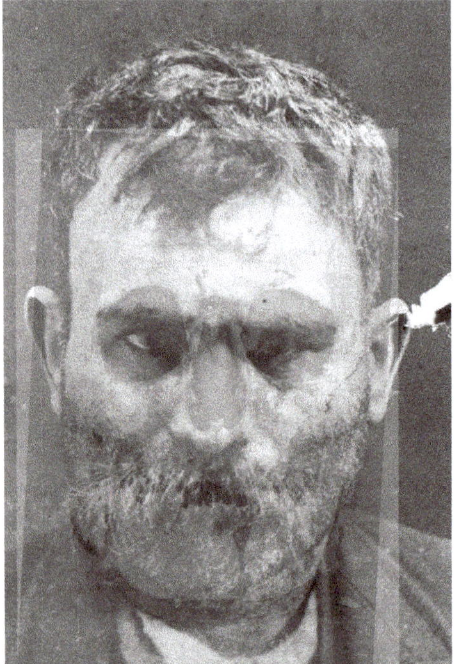

Fig. 6.10: Attempts to resize images of other deaths marks showed none was a reasonable fit. Death masks of executed prisoners Fatta Chand and Alfred Archer are shown, and a facial photo of Antonio Piccone. Victorian Institute of Forensic Medicine.

Chapter 7
The forensic pathology

David Ranson

A combined team approach of researchers was needed to examine Ned Kelly's remains – and those remains provided clues as to how his body was treated after death.

It is fairly well known that in a coroner's death investigation it is the forensic pathologist who examines the body and prepares an autopsy report, setting out the medical findings including the cause – and often the mechanism – of death. What is less often realised is that much of the forensic pathology information is used to both confirm the identity of the individual and to assist in the reconstruction of the circumstances surrounding the death.

This was certainly the case in our examination and identification of the remains of Ned Kelly at the Victorian Institute of Forensic Medicine.

Forensic pathologists cannot do this work alone. Depending on the nature of the case and the condition of the body, the forensic pathologist will put together a forensic team that includes a range of other medical and scientific specialists, whose skills are essential to solving the puzzle of how the death came to occur. When the death investigation involves skeletal remains, for instance, as occurred with Ned Kelly's remains, forensic anthropologists, odontologists and radiologists are critical to the work of the team, with molecular biologists and other laboratory-based forensic scientists analysing samples collected from the body (see Chapters 2, 3, 5, 6, 9).

With such team-based investigations, coordination of the specialists is a critical task for the forensic pathologist. Careful planning is required to ensure that the examination process carried out by one specialist does not compromise the work of others.

Mixing scientific and historical research

Where the investigation involves old or ancient remains, again as in the case of Ned Kelly's remains, the medical and scientific work is often made more difficult by the poor preservation and condition of the bones. In this situation a detailed historical investigation is essential. The meticulous review of relevant historical materials from the time around the death, including the past medical history of the individual, together with information on the way in which the body was handled after death and subsequently disposed of, can be of great help in interpreting the medical findings.

In turn, the medical findings can often help to resolve uncertainties and inconsistencies in the historical records.

Although the historical records clearly show how he died, the Ned Kelly story still contained many uncertainties. There remain several controversial issues with a range of views expressed by different researchers regarding whether an autopsy was performed and what happened to his body (and body parts such as his skull), for instance. The exhumation of the remains of executed prisoners from the Old Melbourne Gaol and their subsequent handling and later reinterment at Pentridge Prison further complicated the forensic identification of individual remains together with skeletal injury determination and interpretation (see Chapter 6).

Many of these questions were interlinked since the historical records described Ned Kelly as suffering injuries in the course of his arrest; if those injuries were to be found in his skeletal remains, this would provide further corroboration of their identity.

From the forensic pathology viewpoint it was clear that the skeletal remains revealed evidence of prior injuries, and indeed evidence of chronic impaired healing of bone around body regions where the historical record indicated that Ned Kelly was wounded (see Chapters 5, 17). The forensic pathology examination was able to extract two projectiles from inside one of the leg bones (without damaging the bones).

The correlation of the injury with the historical record provided important identification evidence which, together with the results of other investigations including the DNA analysis, provided a high degree of certainty regarding the identification of the remains as being those of Ned Kelly.

The question of autopsy

Another question addressed by the forensic pathology examination was the contested issue of whether or not the body of Ned Kelly was subject to an autopsy following his execution.

This requires some explanation of what an autopsy might have involved in that era. Despite the enormous advances in medical science over the last century, the physical dissection processes involved in carrying out an autopsy have changed little. The major cavities of the body are opened and their contents, the internal organs, removed and individually dissected.

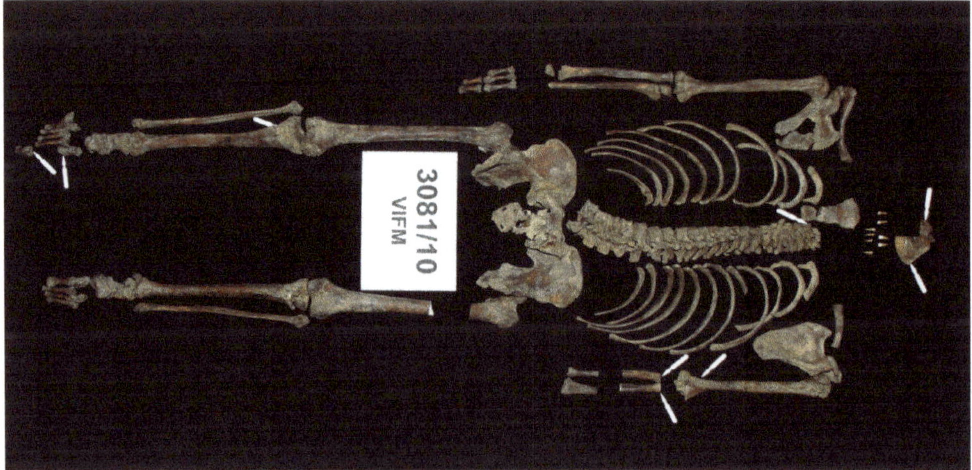

Fig. 7.1: Ned Kelly's skeletal remains. Victorian Institute of Forensic Medicine.

Gaining access to the body cavities involves a skin incision that runs from the skin over the larynx (the 'adam's apple'), down the front of the chest, down the front of the abdomen around the umbilicus (belly button) to just above the pubis. Such an incision allows easy access to the bowel, stomach, liver and spleen. In order to gain access to the heart and lungs, traditionally the front of the ribs are removed with a saw, shears or cartilage knife.

As Fig. 7.1 shows, there is no evidence that any of the bony portions of the ribs in the remains were divided by a saw, shears or knife. This does not completely exclude the possibility that an examination of the internal organs was performed since it is possible to remove the heart and lungs at autopsy, without cutting the bone of the ribs, especially if the subject is a child or a young adult. In such a procedure, the soft cartilaginous parts of the ribs adjacent to the sternum (breastbone) in a young person are able to be divided with a knife. Unfortunately, most of these soft cartilaginous tissues decompose after death and cannot be examined in this type of remains. So it is not possible to say whether the chest of Ned Kelly was opened as part of an autopsy.

Although most of an autopsy is limited to a dissection of soft tissues, which are obviously absent in skeletal remains, some bones are regularly divided during an autopsy. In particular, the bones of the skull are sawn in order to open the cranial cavity and examine the brain. In the case of the remains of Ned Kelly, however, most of the skull – in particular the cranial vault or domed part – was absent. This is the piece of the skull that is usually sawn in order to expose the brain.

Sawing of the bones

Examination of the skeletal remains of Ned Kelly did show areas of sawing of bone. A small portion of the base of the skull was present, and showed saw marks. This portion of skull was the part that includes the foramen magnum, which is the hole in the base of the skull through which the spinal cord exits (see Fig. 7.2).

Not only had this portion of bone been sawn away from the skull, but further saw marks in the adjacent cervical vertebrae or bones of the neck provide a possible explanation for why this portion of the skull bone was removed after death (see Fig. 7.3).

The historical records in relation to executions indicate that an examination of the neck was often performed to assess what had happened to the body as a result of the judicial hanging. To carry out such a dissection, the doctor would position the body face down with the upper portion of the chest lifted above the operating table using a block or small raised plinth. With the head now flexed slightly forwards and the skin and musculature over the back of the neck and the base of the head dissected free from the underlying bone, a handsaw would be used to open the back of the upper neck vertebrae and the back of the base of the skull adjacent to the foramen magnum, in order to inspect the spinal cord as it exits the skull.

Such an approach would also permit an examination of the various ligaments that join the vertebrae and assessment of any damage to the vertebrae at the base of the skull. With the body in this position, the upper portion of the neck lies deeper than the back of the skull and the top of the back of the trunk. As a result, when a straight saw is used to cut through the base of the skull and other vertebrae some of the adjacent structures, including the lower

Fig. 7.2: The portion of occipital bone from Ned Kelly's skull that was present with his skeletal remains. The posterior part of the foramen magnum (the hole at the base of the skull through which the spinal cord connects with the brain) is at the bottom right of the portion. The straight left hand edge of the portion is a saw mark with a transverse saw mark visible at the left top of the portion. These two saw marks meet at a right angle forming the left upper corner of the portion. Victorian Institute of Forensic Medicine.

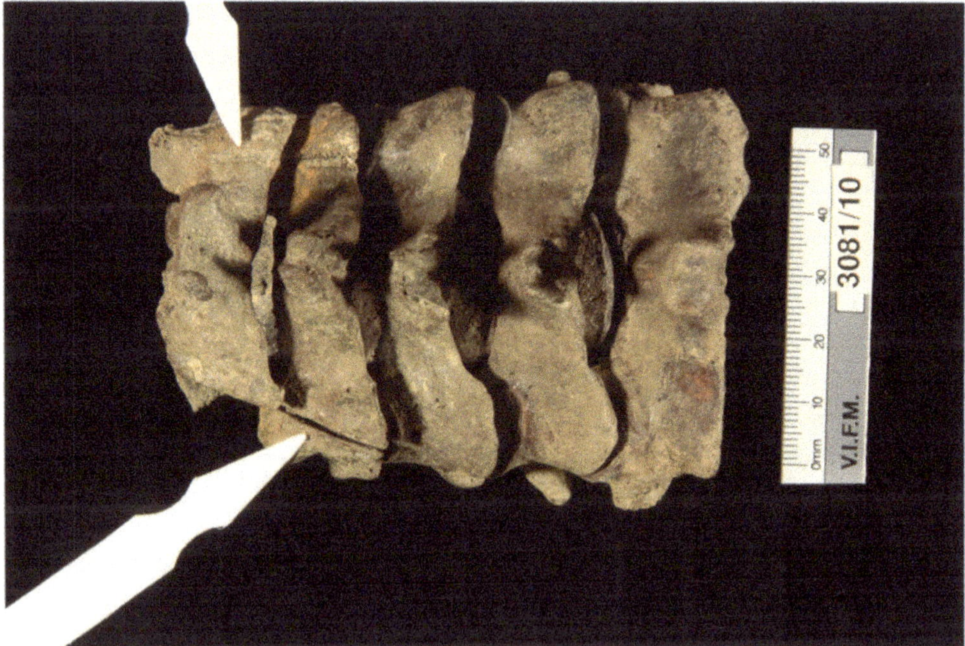

Fig. 7.3: The vertebrae or neck bones, showing old saw marks. Victorian Institute of Forensic Medicine.

cervical vertebrae, may be struck by the saw. The vertebral saw cuts seen in the bone of the lower cervical spine vertebrae in the remains of Ned Kelly are precisely the type of adjacent injury that would be expected if this procedure to examine the effects of the judicial hanging on the missing upper vertebrae had been undertaken (see Chapter 11).

Limited post mortem did occur

While it is not possible to say from the state of the remains that a complete autopsy of Ned Kelly was undertaken, we can say at least that a limited post mortem dissection of the back of his neck occurred.

There is another important outcome of this examination of the neck being undertaken. The dissection of the soft tissues of the neck and the removal of the portions of bone at the base of the skull, as well as portions of the upper cervical vertebrae, would have made for much easier removal of the head from the body by subsequent dissection (or by the processes of decomposition). While speculative, this may provide part of the explanation for why the majority of the skull is absent from the remains of Ned Kelly's body.

Chapter 8
Forensic 3D facial reconstruction

Ronn Taylor

As the forensic sculptor (artist) to the Victorian Institute of Forensic Medicine for the past 22 years I was invited to perform a facial reconstruction on the skull provided by Tom Baxter (see Fig. 8.1). A three-dimensional reconstruction of the face from a skull is carried out to assist in the identification of human remains when more reliable methods fail or are impossible to perform; it is often shown in movies and television.

The idea that our bones give us much of our form and shape is not a new one. The Roman physician and philosopher Galen wrote in the second century AD, 'As poles to tents and walls to houses so are bones to all living creatures, for other features take their form from them and change with them'.

Facial reconstruction from a skull involves seven separate stages:

1 data collection;
2 repair and duplication of the original skull;
3 mounting of the replica skull on a turntable;
4 positioning of soft tissue depth markers and placement of the eyes in the orbits;
5 layer by layer of muscle build-up;
6 fleshing out (outer skin);
7 final details.

Data collection

The forensic sculptor relies on direct communication (as in this case) or written reports from police crime scene investigators, pathologists, anthropologists and forensic odontologists who provide the following information about the age, sex, race and build of the person to give some vital initial information.

Repair and duplication of the skull

Although the reconstruction may be carried out directly on the skull (as advised by Caldwell 1986), it is usually preferable to duplicate the skull in plaster once it has been cleaned and any missing pieces repaired. When I received the 'Baxter skull' it was damaged and some bone and teeth were missing. Using dental modelling wax, the skull was repaired and the missing teeth restored. It was important to restore the teeth in their natural position because the lack of teeth alters the lip support and facial posture. A split mould technique is used in combination with dental alginate impression material to reproduce the original skull (Taylor and Angel 1998).

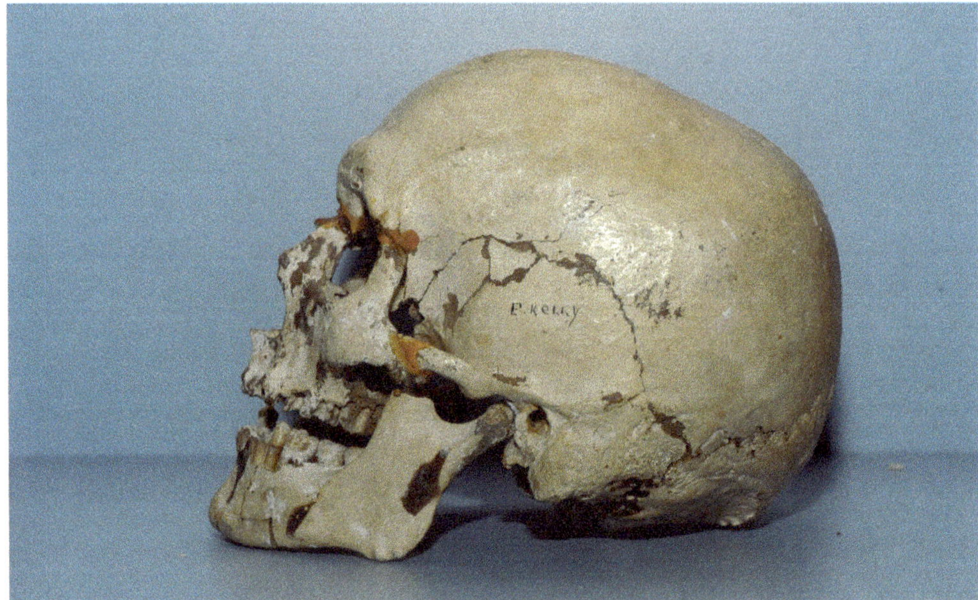

Fig. 8.1: The skull provided by Tom Baxter on 11 November 2009. Victorian Institute of Forensic Medicine.

Mounting the replica skull on a turntable

The replicated skull is mounted via a metal bracket onto a turntable in the Frankfort plane horizontally (see Fig. 8.2) and the skull midline in the vertical position. Sculpting on the replicated skull ensures that no damage occurs to the original and the artist still has the original for reference (Prag and Neave 1997; Taylor and Angel 1998; Wilkinson 2004).

Positioning of soft tissue depth markers and placement of eyes in the orbits

The artist uses soft tissue depth measurements to ultimately form the face (see Fig. 8.3). These are statistical averages of tissue depths which, historically, were obtained by pins being inserted into the faces of cadavers until contact was made with the bone (Miyasaka 1999). On the duplicated skull these measurements are shown by using thin wooden pegs that are placed by drilling holes at right angles into the casting at the given craniometric landmarks, then cut to the required length (Prag and Neave 1997; Taylor and Angel 1998; Wilkinson 2004).

Layer by layer of muscle build-up

In general terms, the muscles of the face are grouped into those of mastication (chewing) and those of facial expression. As facial reconstruction is based on the armature or framework of the skull, what makes the process difficult is that most of the muscles of facial expression have only one attachment to the skull. Some do not attach to bone at all, and others are very small. The main muscles for mastication, the masseter and the temporalis, have clear attachment origin and insertion points, however.

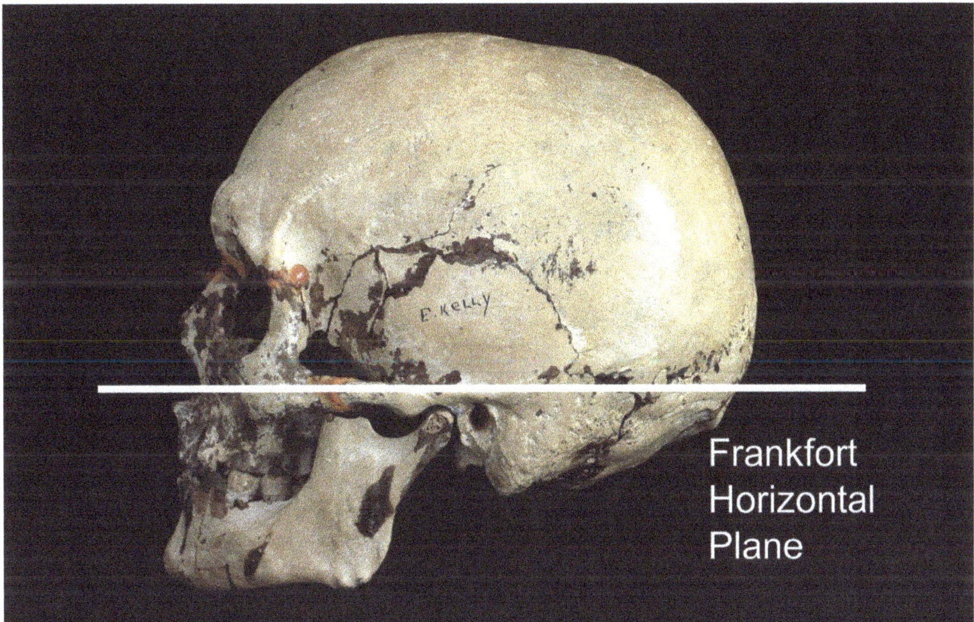

Fig. 8.2: Frankfort plane. Victorian Institute of Forensic Medicine.

Fleshing out (outer skin)

A thin layer of clay like a pizza base is next used to cover the underlying muscles. The problem soft tissue areas such as the ear, mouth and tip of nose, which have wide variation, constitute the greatest challenge to the forensic artist.

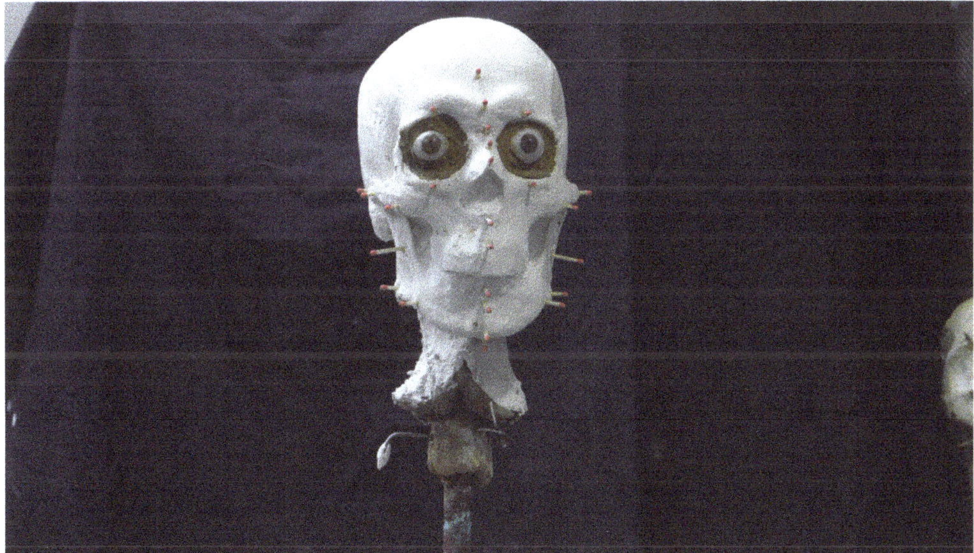

Fig. 8.3: Mounted plaster skull with soft tissue markers and eyes in place.

Final details

The final detailing is added manually, as in this case with the skull thought to be Ned Kelly's. The reconstruction can be scanned and enhanced digitally with the addition of hair, eyebrows, clothing and jewellery etc.

The final result can then be compared with photographs or paintings, for instance, to determine how close a match it may be (see Figs 8.8, 8.9).

Facial build-up is a gradual process, as shown in Figs 8.4 to 8.7.

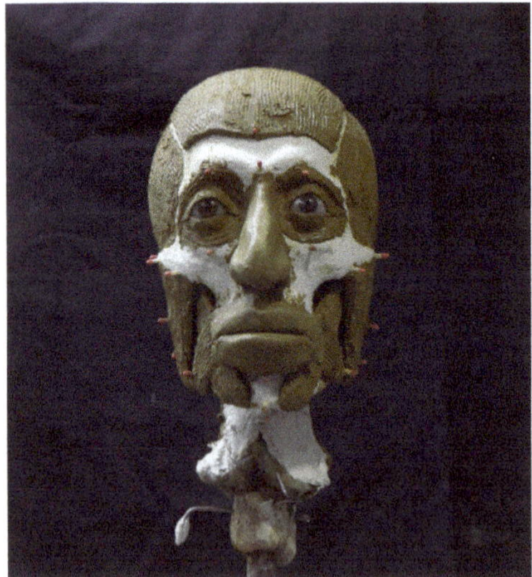

Fig. 8.4: Gradual build-up of facial reconstruction.

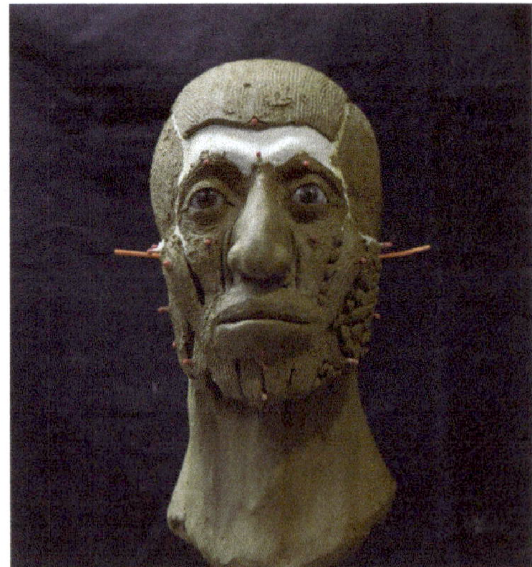

Fig. 8.5: Gradual build-up of facial reconstruction.

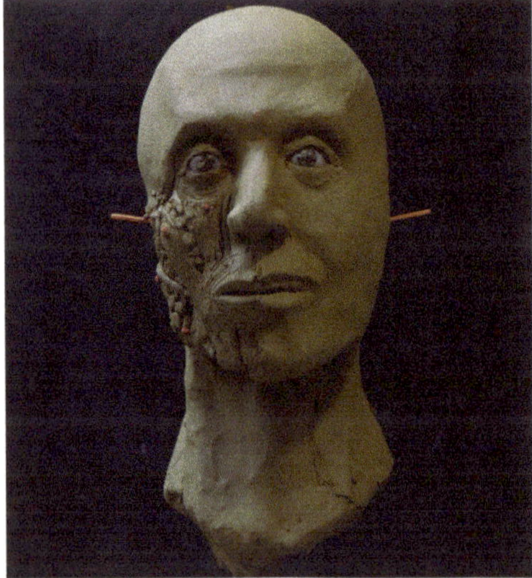

Fig. 8.6: Gradual build-up of facial reconstruction.

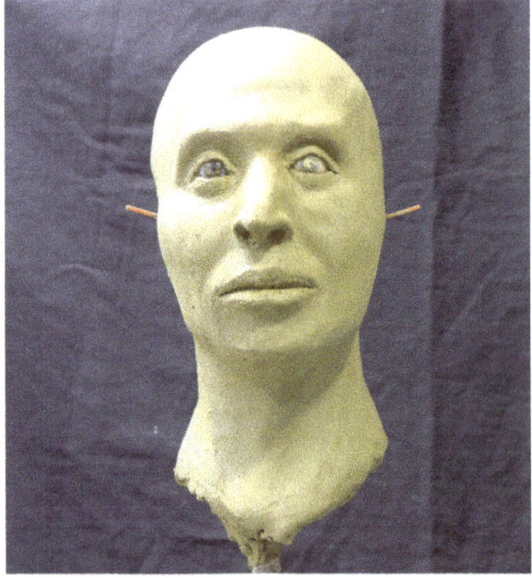

Fig. 8.7: Gradual build-up of facial reconstruction.

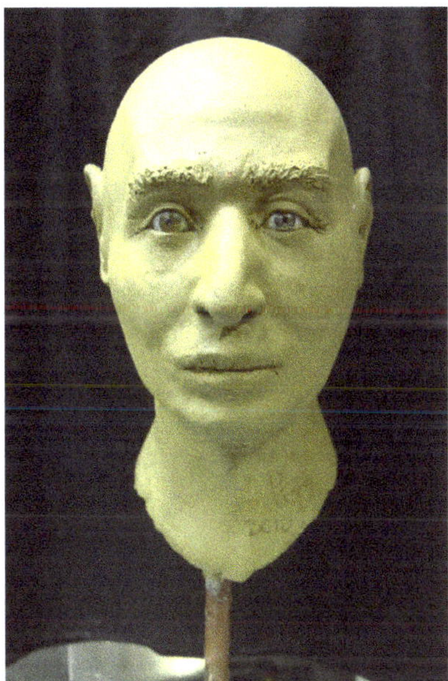

Fig. 8.8: The completed head.

Fig. 8.9: Ned Kelly portrait by William Burman, July 1880. State Library of Victoria.

References

Caldwell MC (1986) New questions and some answers on facial reconstruction techniques. In *Forensic Osteology: Advances in the Identification of Human Remains* (ed. KJ Reichs) pp. 229–255. Charles C. Thomas, Springfield.

Miyasaka S (1999) Forensic. *Scientific Review (Singapore)* **11**(1), 55.

Prag J, Neave R (1997) *Making Faces: Using Forensic and Archaeological Evidence.* British Museum, London.

Taylor RG, Angel C (1998) Facial reconstruction and approximation. In *Craniofacial Identification in Forensic Medicine* (eds JG Clement and DL Ranson) pp. 177–185. Arnold, London.

Wilkinson C (2004) *Forensic Facial Reconstruction.* Cambridge University Press, Cambridge.

Chapter 9
Turning to the DNA

Dadna Hartman and Carlos Vullo

Understanding DNA extraction and analysis can be difficult, but is the basis of understanding both how DNA profiling is used, and how accurate it can be for identifying human remains. When a direct DNA profile is not available from the person of interest for comparison purposes, it is possible to use the DNA from a living relative as a reference sample to confirm identity.

DNA analysis can be such an effective tool for forensic human identification that it has been used to identify badly burned and damaged bodies after airline crashes or terrorist explosions. And ancient DNA analysis has been used to suggest the parents of the Egyptian pharaoh Tutankhamen, Akhenaten and mother Nefertiti, who died over 3300 years ago, were in fact closely related.

However, the accuracy of forensic DNA identification, more often used in criminal investigations, depends to a large extent on obtaining good-quality DNA samples to work with – these can sometimes be very difficult to obtain.

The key challenge in identifying Ned Kelly's skeletal remains by DNA analysis was finding DNA samples to compare them with that could be positively said to be his, or were from a known close relative. A specialised team of staff from the Victorian Institute of Forensic Medicine, joined with colleagues from the Argentine Forensic Anthropology Team (EAAF), to undertake the identification of Ned Kelly's remains, following special funding provided by the then Victorian Attorney-General, Rob Hulls. The EAAF was involved because of its experience in human identification in different countries for the past 25 years. Its staff have developed a dedicated DNA forensic laboratory specialising in the analysis of

EAAF

The Argentine Forensic Anthropology Team (EAAF), established in 1984, is a non-governmental, not-for-profit, scientific organisation investigating human rights violations worldwide through the use of forensic science. Its original charter was to investigate the cases of at least 9000 disappeared people in Argentina, and it has found and identified the remains of 577 persons disappeared during the 1976–1983 military dictatorship there. Today the team continues its work in more than 30 countries in Latin America, Africa, Asia and Europe, often at the request of local and international human rights organisations or local and international judicial bodies, such as Truth Commissions.

http://www.eaaf.org/

several nuclear genetic markers as well as mitochondrial DNA (mtDNA) analysis (see p. 90) from aged human skeletal remains, and it was clearly the best partner to turn to.

Conducting the DNA analysis

Following the anthropological analysis, a bone and/or tooth sample was collected from each case of skeletal remains recovered from Pentridge Prison for DNA analysis. Samples were sent to the EAAF DNA forensic laboratory in Argentina, to undergo mtDNA analysis. This type of analysis is a useful forensic tool, because it can yield useful DNA profiles from degraded samples types with poorly preserved nuclear DNA (nDNA), and due to its mode of inheritance. MtDNA analysis can be used to trace maternal relatives (or lineages) in families across many generations.

Given the antiquity and condition of skeletal remains recovered, and the 130 years that had passed since Ned Kelly's execution, mtDNA analysis was chosen as the best DNA analysis tool to help in the inclusion or exclusion of remains as being those of Ned Kelly. The bones that were examined for DNA included teeth, collar bones, leg and arm bones and skull fragments.

Ensuring that there was no DNA contamination from outside sources was very important, so the DNA examinations were carried out in a specially designed laboratory. Covered cabinets that restrict air flow from outside by directing air flow and using filters, known as laminar flow cabinets, were used in every step of DNA extraction and examination. All plastic materials used were DNA-free, based on their certification or following UV irradiation. All staff working on the examination had to wear protective equipment, including facial masks, hair coverings, single-use laboratory gowns and sterile gloves.

Any possible external DNA was removed from workbenches using a bleach solution and UV light irradiation. And finally, all the tools used to work with the bone samples, like tweezers, forceps and the power tools used for carving, cutting and sanding, were also subjected to UV irradiation (see 'DNA workflow', p. 90).

DNA extraction

The process for extracting DNA is quite complex and requires blank indicator samples to be analysed side by side with each sample (see 'Blank samples', p. 92). This is a safety check to ensure that there has been no inadvertent DNA contamination. If any of the blank samples provided a DNA reading, there would clearly have been some contamination and the results relating to the samples would be in doubt.

Teeth and bone remains were first cleaned with a brush and neutral soap then, depending on the type of skeletal sample, a sterile scalpel or a power tool with a sanding function was used to remove any dirt from the sample. Bone remains also had to be cut using a handsaw or a power sanding disk. After this, all bones and teeth processing was carried out in DNA-free laminar flow cabinets.

Prior to extracting the DNA, the samples were washed in distilled water, soaked in bleach to remove any surface (contaminating) DNA, washed again several times with distilled water, then had a final wash in ethanol before being allowed to air-dry in the laminar-flow cabinets. They were then exposed to UV light.

The teeth or bone remains were next ground to a powder using a freezer mill (grinder), and other substances that were present in the bone sample together with the DNA (e.g. calcium) were removed by a process known as decalcification. This was done to maximise the recovery yields of DNA from each sample. Decalcification is done by placing the samples in a solution of acid (ethylenediaminetetraacetic acid), with constant stirring. The decalcified bone powder is placed into a centrifuge, and the resultant material is then ready for DNA extraction.

This is done by first disrupting the cells – leading to the release of the cellular content including DNA molecules – then the DNA is recovered using a DNA-binding column (known as silica binding column kits). This binds the DNA molecules to a column, allowing all other cellular components to be washed away. The bound DNA is then released from the column as a purified DNA solution. This is like fishing with bait that binds to only one type of fish, and leaves everything else behind in the water.

Once extracted, it is important to know the quality of the DNA recovered. This is because the techniques used to analyse DNA have particular thresholds for the minimum amounts and size of DNA required. If there is too little DNA or if the DNA is too degraded, then analysis cannot be undertaken.

The quality of the DNA extracts was checked using commercially available kits to specifically quantify human DNA, and those DNA extracts deemed to have sufficient good-quality DNA were subjected to mtDNA analysis.

However, obtaining mtDNA profile information for the Pentridge cases alone was not sufficient for the identification of Ned Kelly's remains, as a comparison sample from a maternal relative was needed for the researchers to undertake kinship comparisons.

Searching for a maternal descendant

So, having established that mtDNA was available from the different skeletons, the search for a maternal descendant of Ned Kelly began, to find a person who would have the same mtDNA profile as Ned himself. Fortunately Ned Kelly's mother's great-great-grandson in a direct female line was located and he agreed to assist the researchers. Mr Leigh Olver is a teacher in Melbourne and his lineage was confirmed through records held at the Victorian Registry of Births, Deaths and Marriages.

He became known in the study as 'MRNK' (maternal relative of Ned Kelly).

Once a biological sample from Mr Olver was taken, it was subjected to the same mtDNA analysis as the Pentridge samples. The researchers now had an mtDNA profile for the maternal lineage of Ned Kelly and could compare it to the mtDNA profiles obtained from the Pentridge remains.

Using specialised computer software capable of aligning mtDNA sequencing data, the mtDNA profiles of Mr Olver and those from the Pentridge remains were compared to determine if there was a match (see p. 245).

Trying to get nuclear DNA samples

In addition to mtDNA analysis, the team had to consider the possibility – however remote – of conducting nuclear DNA analysis, obtaining the DNA from the nucleus of a cell of both the skeletal remains from Pentridge and any Ned Kelly artefacts that had his biological

Mitochondrial DNA (mtDNA) analysis

There are two types of DNA that can be recovered from most cells in the human body: nuclear DNA (nDNA) from the nucleus of a cell, and mitochondrial DNA (mtDNA). As the name suggests, mtDNA resides in the mitochondria, which can occupy the non-nucleus parts of the cell. Using an egg as an example, nDNA would be found in the yolk and mtDNA in the white.

Mitochondrial DNA has several characteristics that set it apart from nuclear DNA. It is much smaller in size and circular in shape, and there are 100-fold more copies of mtDNA per cell. Importantly, mtDNA has a maternal mode of inheritance. That is, a mother will pass on her mtDNA type to all her offspring, only the daughters will pass on the same mtDNA type to their children and so on. Hence, in a given family line, all the maternal relatives should share exactly the same mtDNA type.

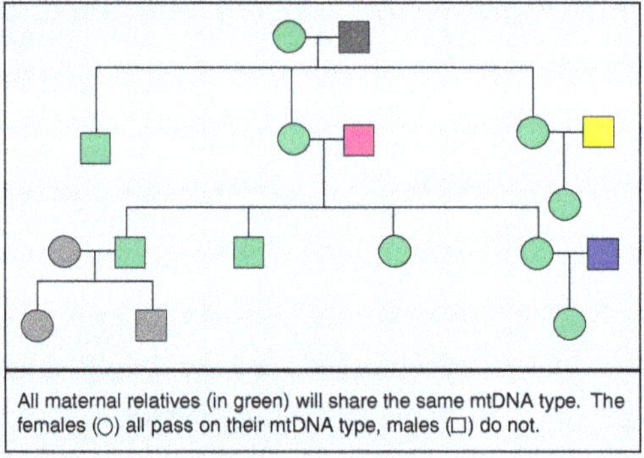

Mitochondrial DNA: all maternal relatives (in green) will share the same mtDNA type. The females (circle) all pass on their mtDNA type, males (square) do not.

DNA workflow

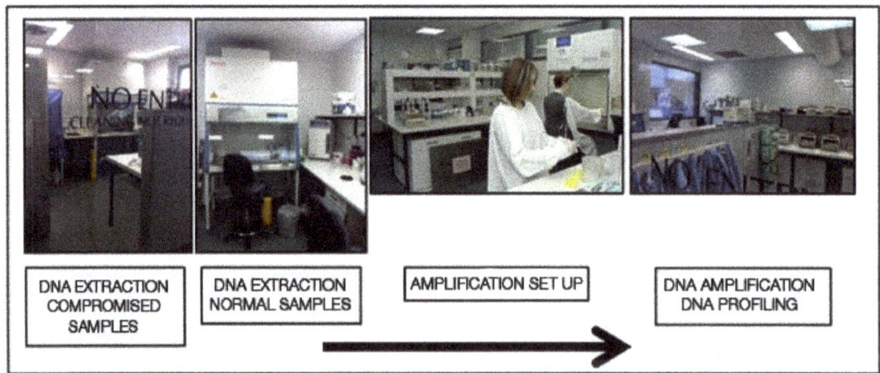

The steps in the DNA analysis workflow are unidirectional – each step in the processing of a sample is performed at a different time and location to minimise contamination events. For example, samples that are in the amplification stage are not processed in the proximity of samples undergoing DNA extraction.

material on them. If this was possible, it would be a primary form of identification. The only likely sources of Ned's DNA, however, were personal items known to have belonged to Ned that had been in direct contact with him (enabling the transfer of DNA from his person to the item). The historical researchers had already located two potential sources for comparative reference material for DNA analysis, objects that may have been contaminated with Ned Kelly's blood while he was still alive. These included a sash believed to have been worn by Ned Kelly at the time of the shoot-out at Glenrowan, which had been on display at the museum in Benalla, and his boots – one of which had been filled with blood after he was shot in the foot.

Unfortunately, samples collected from these items failed to yield suitable nDNA profiles for comparison. In addition, nDNA profiles could not obtained from any of the Pentridge samples, which was not too surprising given their age and the high risk of degradation.

Making sense of the DNA data

Given the nature of mtDNA analysis (where some individuals may be found to share the same haplotype), it was necessary to attempt to profile all the individuals recovered from the Pentridge site. At least one representative sample from each of the remains (mostly collar bones) from all three pits was selected for DNA analysis. The best skeletal elements for DNA profiling are actually teeth and the compact portion of long bones (femur and tibia) of the legs, or the upper arm (humerus). This is because they are thicker and contain more bone material for analysis in comparison to other bones. Teeth (particularly molars) are a very good source of DNA, as the enamel protects the inner pulp chamber which is a good source of DNA. By comparison, the spongy component of bone produces poor results (Edson *et al.* 2004).

It is worth pointing out that most of the bones analysed in this study were collar bones, which often give poor results. They were chosen as they were the commonest bone present in the Pentridge cases – which did not all have complete skeletal remains, and in many cases were mixed. Similar to the nDNA analysis that showed very poor or no results, due to the degraded nature of the DNA because of the age of the samples and their burial conditions, degradation was high for the mtDNA analysis. It was therefore necessary to perform several amplifications and sequencing runs of the mtDNA recovered from the samples, to effectively increase the amount of DNA able to be analysed.

As the mtDNA analysis of the initial samples progressed, additional skeletal elements were also analysed. This was partly to compensate for those collar bone samples that failed to yield mtDNA for analysis, as well as to confirm the mtDNA profile of several cases by analysing multiple samples. In total, 64 bone samples – including clavicles, teeth, femurs, cranium, tibia and ulna – were subjected to mtDNA analysis. Not surprisingly, the success rate in obtaining mtDNA sequences for analysis was dependent on the bone type analysed and on their preservation (see Fig. 9.1). Many bone samples were around 100 years since death, and poorly preserved.

An mtDNA profile was obtained for 29 of the 36 cases from Pentridge (see Table Appl.1). Of these, 27 could be classified into known mitochondrial haplogroups, while the other mtDNA profiles remained unclassified. Of the remaining nine Pentridge cases, seven

Blank samples

Due to the sensitive nature of DNA testing and the added risks that samples may become contaminated by external DNA sources, it is important to ensure that the DNA profile being examined is indeed from the sample and not from an external source. To this end, scientists conducting DNA testing include what are known as blank samples as part of their testing regime. These include the extraction and amplification blank samples.

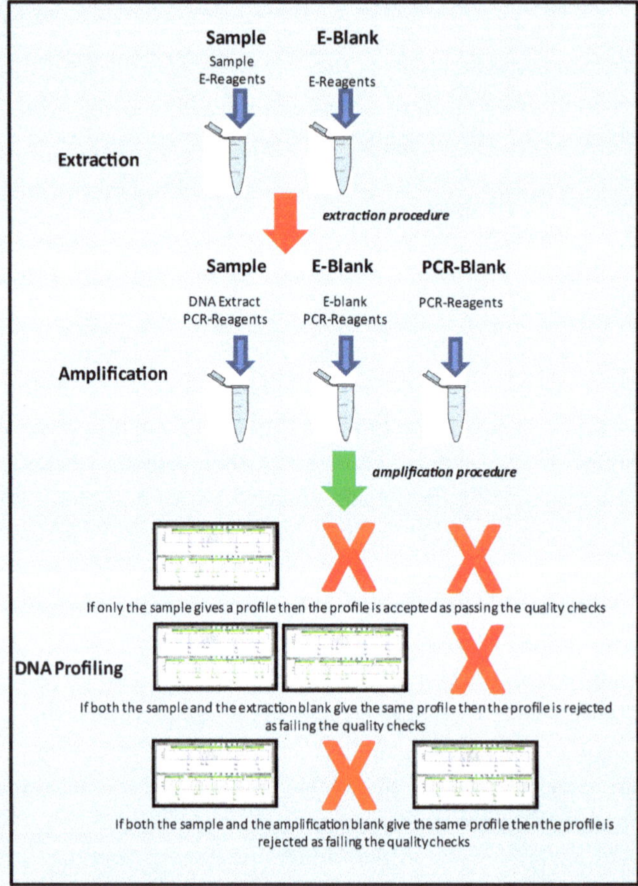

If the sample yields a DNA profile that is also present in either or both blanks, then it is highly likely that the profile is the result of contamination, rather than from the sample itself. The contamination can come from other samples, reagents or equipment used, or even from a scientist in the laboratory.

Another quality step undertaken as part of the DNA analysis process is the use of a positive control sample. This is a sample, whose DNA profile is known, that is put through the same steps as the sample to be tested. This will tell the scientist if the method is working properly. For example, if the sample and blanks all fail to give a DNA profile but the positive control sample gives the expected DNA profile – then it is not due to a process failure that the sample has not returned a DNA profile. In this instance, it is more likely that there was not sufficient DNA present in the sample for a DNA profile to be generated.

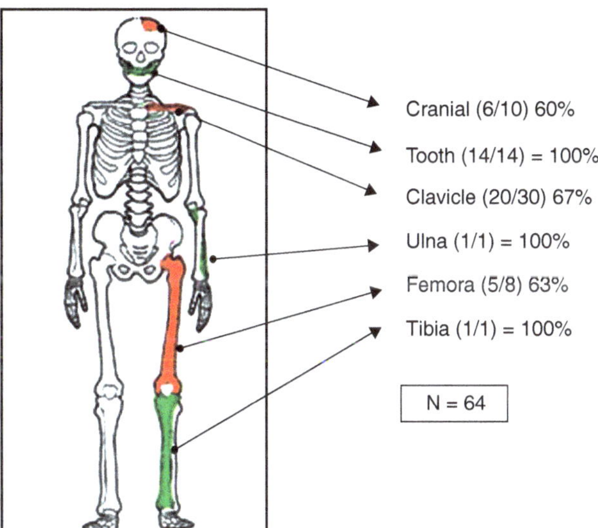

Fig. 9.1: mtDNA typing success rate versus bone type for the 64 Pentridge samples analysed. Blau *et al.* 2014.

failed to yield any reportable mtDNA for analysis and two gave only a partial mtDNA profile, all due to highly degraded DNA.

DNA analysis also showed that 23 of the 27 cases with full (HV1/2) profiles belonged to the most common haplogroups in Europe and three belonged to Asian-Oceania haplogroups, with one being unclassified. Further comparisons indicated that the distribution of European haplogroups among the Pentridge cases was similar to that for western Europe (see Table Appl.2) (Mitomap). This was not entirely surprising, given the historical information about the ancestry of the inmates at the Melbourne Gaol during the 1800s.

More importantly, though, only one of the Pentridge cases – Pen14, or more strictly speaking case number 3081/10 – matched the mtDNA haplotype of MRNK. The mtDNA haplotype also corresponded to the haplogroup J1c, which is predominantly European (Finnilä *et al.* 2001; Richards *et al.* 2000, 2002). To confirm the mtDNA profile of 3081/10, additional samples of the skull fragments and bones from the arm and leg (ulna and femur) were also subjected to mtDNA analysis. All returned the same mtDNA haplotype.

Clearly, 3081/10 was related to MRNK!

In plain speaking, the skeletal remains of one prisoner could be related in a maternal line to Leigh Olver, a descendant of Ned Kelly's mother. In scientific speaking: 3081/10 could not be discounted as being Ned Kelly.

To be thorough, for the seven remaining Pentridge cases for which an mtDNA profile was not obtained, anthropological and odontological examinations were used to exclude the remains as being those of Ned Kelly. Researchers at the Institute could now say they were increasingly confident that the body of Ned Kelly had been identified (Blau *et al.* 2014).

However, there was still the issue of the skull. Additional mtDNA analysis of the only cranial fragment found with the skeletal remains of case 3081/10 (a fragment from the back of the head) established that the sample matched with the reference sample provided by the maternal descendant of Kelly. No such match was made to the skull that for so many years was thought to be Ned Kelly's (the 'Baxter skull'), so it was conclusively shown not to be his.

If the remainder of Ned Kelly's skull is ever located, testing it may be a little easier, as the arduous preliminary work for analysis and identification has been done.

References

Blau S, Catelli L, Garrone F, Hartman D, Romanini C, Romero M, Vullo C (2014) The contributions of anthropology and mitochondrial DNA analysis to the identification of the human skeletal remains of the Australian outlaw Edward 'Ned' Kelly. *Forensic Science International* **240**, e11–e21. http://dx.doi.org/10.1016/j.forsciint.2014.04.009.

Edson SM, Ross JP, Coble MD, Parson TJ, Barritt SM (2004) Naming the dead: confronting the realities of rapid identification of degraded skeletal remains. *Forensic Science Review* **16**, 63–90.

Finnilä S, Lehtonen MS, Majamaa K (2001) Phylogenetic network for European mtDNA. *American Journal of Human Genetics* **68**, 1475–1484. doi:10.1086/320591.

Mitomap. http://www.mitomap.org/bin/view.pl/MITOMAP/HaplogroupFreqs.

Richards M, Macaulay V, Torroni A, Bandelt HJ (2002) In search of geographical patterns in European mitochondrial DNA. *American Journal of Human Genetics* **71**, 1168–1174. doi:10.1086/342930.

Richards M, Macaulay V, Hickey E, Vega E, Sykes B, Guida V, Rengo Ch, Sellitto D, Cruciani F, Kivisild T, Villems R, Thomas M, Rychkov S, Rychkov O, Rychkov Y, Gölge M, Dimitrov D, Hill E, Bradley D, Romano V, Calí F, Vona G, Demaine A, Papiha S, Triantaphyllidis C, Stefanescu G, Hatina J, Belledi M, Di Rienzo A, Novelletto A, Oppenheim A, Nørby S, Al-Zaheri N, Santachiara-Benerecetti S, Scozzari R, Torroni A, Bandelt H (2000) Tracing European founder lineages in the Near Eastern mtDNA pool. *American Journal of Human Genetics* **67**, 1251–1276. doi:10.1086/321197.

Chapter 10
Looking after Ned in the mortuary

Jodie Leditschke

As interest in Ned Kelly was high and there was a history of people 'souveniring' his remains, the Victorian Institute of Forensic Medicine had to keep his remains under lock and key in its mortuary and limit access to them. When the news of the Institute findings was released, though, a tough decision on existing protocol had to be made – whether to allow his remains to be viewed by the media.

In May 2009, 21 coffins containing the human remains of executed prisoners exhumed from the site of Pentridge Prison were admitted to the Victorian Institute of Forensic Medicine's mortuary. At the time we believed we would store the coffins for two months until a decision was made on where they should be reinterred. That turned out to be a very conservative estimate!

On an average day, the VIFM mortuary houses approximately 140 to 160 deceased persons, of which about 50 are partial or complete skeletons. It is a policy, and a firm cultural belief, that all deceased persons or human remains are treated in the same manner and with the same respect, regardless of whether they are a vagrant from the street with no family or a famous sportsperson who has met an untimely death. Everyone is examined, attempts are made to identify them, and they remain covered and secured in our fridge with restricted access.

It was the same with all the human remains from Pentridge Prison.

Each set of human remains was first scanned using a multi-slice CT scanner then, under the supervision of the anthropologist, the remains were scrubbed and cleaned by the mortuary staff (see Chapter 5). Most of the remains had been covered in clay. Then they were laid out anatomically, ready to be examined and photographed.

Once the pathologists and anthropologists completed their examination, the odontologist examined the dentition if it was present (see Chapter 6). Finally, sampling of the bones for DNA took place (see Chapter 8). Each sample was labelled, security sealed and transferred to the laboratory. Each step was carefully documented and tracked, in the same way we treat all our unidentified human remains. At all times particular care was taken to ensure the remains were well labelled and not inadvertently mixed-up, although many coffins already contained multiple individuals. The method and process was similar to the way we manage mass fatalities at the Institute.

It was rumoured that Ned Kelly was possibly among the remains but we had to wait for the results of DNA testing.

Buying a safe for the skull

On 11 November 2009, Tom Baxter visited the Institute and delivered a skull he firmly believed belonged to Ned Kelly (see Chapter 1). It was impressed on me that other people might want to souvenir the remains and that I was to ensure its security, so we bought a safe. Tom Baxter ceremoniously handed me an old make-up case containing the skull, and it was admitted to the mortuary with a coronial case number as it was an unidentified skull. It was photographed and scanned as per normal protocol; however, in this instance we placed the skull in a safe to which only the Director and I knew the combination.

Over the next few months, whenever anyone wanted to examine the skull I would retrieve it and ensure it was returned to the safe. Even when the skull was determined not to be that of Ned Kelly it remained in the safe until we could determine with which skeleton from Pentridge Prison it should be placed.

When Ned Kelly was finally identified, then his whole skeleton had to be secured. Once it was suggested that someone may want to souvenir a part of him, it was difficult to treat him the same as everyone else. I locked him in a separate fridge and kept the key.

Granting media access

In the days leading up to the public announcement of the identification of his remains, a decision was made that one outlet representing all the media could view the skeleton. Never before have we allowed the media into our mortuary to view a deceased person's remains: the decision to do this was debated and not taken lightly, as it was against our policies and all previous practice. It was finally decided that Ned Kelly's remains fell into a category of being historical and that he was such an Australian icon that in a sense he was part of our common heritage. It was thus in the public interest for one outlet of the media to view the remains, share the footage and photographs with all other media so that the public could see for themselves.

Personally, I still am not sure whether I agree with this but respect the decision made by people higher in the chain. It was difficult to go against everything we believed in and put his remains on display. In order to minimise the disrespect to the remains, only one film crew was allowed to enter and their footage and photographs were distributed to all networks.

Many years ago a colleague asked if they could see the body of a famous person who died an untimely death. There was no professional reason for the colleague to view the body so obviously I denied the request. I also made sure the person was stored in a separate secured refrigerator. Deceased persons stored at the VIFM are stored for a very specific purpose, they are not to be viewed out of curiosity. This is very clearly understood by everyone at VIFM, even more so in recent years.

Throughout the next few months, as debate raged over what was to happen with Ned's remains and the media showed increasing interest, I worried that we would lose a part of him or, worse still, someone would steal a part. I trusted my staff and colleagues explicitly but the Institute was never designed for museum-type security.

One day, when a descendant of Ned Kelly's family was viewing his skeleton, a colleague mouthed the words to me 'Where is the other shotgun pellet?' I panicked. There were meant to be two lead pellets with his remains and she said there was only one. I searched the inside of the fridge, underneath the racking of other remains and all other areas of the mortuary. The colleague came out and told me that she had found it lying under the bone of the other leg. I then placed the pellets in the safe.

Ned Kelly was treated differently by us in some ways but was treated the same as all deceased persons in other ways. Each examination and each protective measure was completed so we could ensure he was returned to his proper custodians – the family. After two years he was finally given back to his family. I was relieved.

Recently my son needed to choose a person to depict at the end-of-year 'Night of Notables' school function. He chose Ned Kelly. Regardless of what individuals each think of Ned Kelly the Australian community is still fascinated by him – although I think Ned would have laughed at our efforts to protect him from being stolen, given the many robberies he was responsible for in his time.

Chapter 11
Judicial hanging: the injuries and effects

Stephen Cordner

Despite being practised for many centuries, death by hanging has a long history of often being bungled, which continues today.

Asphyxia, compression of the neck, strangulation, suicidal hanging, judicial hanging, fracture of the cervical spine – these things are jumbled in the average person's mind when they hear of a person dying as a result of hanging – but they are all different, and forensically understanding the differences can be quite important.

The starting point for an understanding of what happens in any sort of hanging is the anatomy of the neck. Briefly, travelling through the neck are several crucial, vital, structures:

- the carotid arteries taking blood from the heart to the head including the brain (supported by the vertebral arteries travelling up within the cervical vertebrae);
- the internal jugular (and other) veins returning blood from the head to the heart;
- the trachea (or windpipe), protected in its upper reaches by the laryngeal cartilages (including the voice box or adam's apple), through which inhaled air reaches the lungs;
- the vagus nerve, a multi-functional nerve travelling from the base of the brain to a wide variety of destinations including the heart;
- the cervical spine, its bony vertebrae and associated extremely tough ligaments, protectively containing within it the soft spinal cord.

These are the structures at risk whenever the neck is damaged and when a person dies from hanging, they die because their neck is compressed or injured by a ligature concentrating the weight of the person's own body on the neck.

However, there are several variables that affect this brutal outcome and the speed with which it occurs. They include:

- the distance of the drop;
- the positioning of the knot (or the metal eye through which a running noose is created);
- the strength of the muscles, tenacity of the ligaments and resilience of the bones of the neck;
- the weight of the individual.

The picture most of us have in our minds when we think of death from hanging has been outlined by Albert Pierrepoint, who from the early 1930s until the mid 1950s followed in his father Henry's and uncle Thomas's footsteps as an English executioner. He wrote:

> The executioner and his assistant wait outside the condemned man's cell, with the chief officer and second officer detailed to conduct the prisoner to the execution chamber. On a signal given by the sheriff, they enter, and the executioner pinions the prisoner's arms behind his back. He is escorted to the drop with one officer on either side. The sheriff, the prison governor and the medical officer enter the execution chamber directly by another door.
>
> The prisoner is placed on the drop at a marked spot so that his feet are directly across the division of the trap doors. The executioner places a white cap over the prisoner's head and puts the noose round his neck, while the assistant pinions his legs. When the executioner sees that all is ready he pulls the lever. The medical officer at once proceeds to the pit and examines the prisoner to ascertain that life is extinct. The shed is then locked and the body hangs for one hour. The inquest is held the same morning (Pierrepoint 1977).

The drop, and the sudden halting of the falling body by the noose at its end, fracturing the cervical spine and damaging the upper spinal cord, resulted in more or less instantaneous death, and made the hanging much more effective. The drop also increased the likelihood that the hanging would in fact result in the death of the prisoner, and quickly.

However, despite the fact that execution by hanging has been practised for many centuries, the earlier means of hanging did not usually incorporate a drop. For example, Henry VIII executed 72 000 of his subjects. Most were led by the executioner up a ladder, and a noose that was tied to a beam between two trees was placed around their neck. The unfortunate person would then be turned off the ladder and simply left to dangle. They would frequently struggle violently before succumbing to unconsciousness and death (Laurence 1926).

Execution needing to be repeated

In this ghastly situation, probably involving a short drop, the death is generally a more prolonged event. If the arrangement of the noose results in the front of the neck being compressed then it might be perhaps 30–60 seconds before unconsciousness occurs, with death following shortly afterwards. If not, and the noose somehow slips so that mainly the side and back of the neck are bearing the weight of the body, then it may take considerably longer for unconsciousness and death to follow – or death might not even occur at all, meaning that the execution would need to be repeated.

When the noose focuses the weight of the body on the front of the neck, it compresses the veins, arteries and airways running through the neck. Thus, the blood supply to and from the brain, and the supply of air to the lungs, are all severely compromised. Stopping the supply of blood to the brain will result in consciousness being lost in under a minute or so at the most. However, many cases of hanged men surviving the experience are on record. The most celebrated of these is probably 'half-hanged' Smith whose pardon arrived after he had

The effects of hanging

From the official record book of hangings at the Melbourne Gaol and Pentridge Prison between 1894 and 1967 (the Sheriff's Particulars of Execution) it would seem that the cause of death in most, but not all, executions at the Melbourne Gaol was fracture or dislocation of the second cervical vertebra (referred to as C2 or the axis vertebra), although vertebrae below this level could also be affected. The medical notes contained expressions such as 'the spine was dislocated' (Frances Knorr's execution in 1894), 'the neck was found to be broken' (Martha Needle's execution in 1894) and 'fracture of the neck' (Alfred Archer's execution in 1898). More detailed anatomical descriptions in the book accompany executions that took place in the early 20th century:

> *The specimen shows that the 2nd cervical vertebra was fractured at the meeting of the body with the arch of one bone and resultant compression of the spinal cord, by the look of the vertebrae (fracture-dislocation) (August Tisler's execution in 1902).*

Although few complete cervical spines were found among the remains examined, we were interested to examine all vertebrae as that might confirm the cause of death. Only four complete cervical spines were found that comprised all seven cervical vertebrae, and there were other cases with almost complete cervical spines, with only the seventh cervical vertebra absent. In the remainder of the boxes the cervical spines were essentially incomplete or absent. Nevertheless, certain conclusions could still be drawn even from these fragmentary remains.

For example, four specimens clearly showed evidence of the 'hangman's' fracture – although this term was not in use at the time, not being coined until the 1950s. They included damage through both sides of the vertebra at the pedicle, an anatomical site located just behind the vertebral body, which is its weakest point. The fracture most commonly involved C2, although in one spine there was a fracture at C3/4.

A fracture may have been present in a further four instances, but it is difficult to be sure because of incomplete preservation. Interestingly, in two of these cases only the front parts of the third and fourth cervical vertebrae survive while vertebrae close to them (the second and sixth) are completely intact and well preserved.

Fiona Leahy and Chris Briggs

been suspended for 15 minutes but who nevertheless recovered (Crook 1926). He must have been fortunate that the noose was not compressing the front of his neck.

While consciousness exists, the poor person hanging may well exert himself against the hanging by writhing and contorting – 'dancing' as it has been called – at the end of the rope. This would be less obvious, though, if the arms and legs were tied. Even after consciousness is lost, however, there may be twitching or fitting, quite indistinguishable to those watching from 'dancing' (almost certainly including family in these older times) from a conscious individual.

Indeed, relatives and friends would sometimes rush forward to pull on the legs to lessen the time taken to die, or even to carry off the victim, sometimes competing with groups wanting the body for anatomical dissection.

The wider toll of execution by hanging

Undoubtedly executions exacted a significant toll, not just on the victim, but also on those in authority.

The Victorian executioners and flagellators (charged with whipping prisoners) were men of dubious character even before the nature of their responsibilities took its toll. The *Argus* reported that:

> *The effect of such an office on an individual may be learned from the evidence of warders who have known [hangman] Gately for years. They all say that his conduct now is greatly worse than when he was an ordinary criminal. Let this be record to his credit – he could be degraded and made worse by the influences of his debasing office.*

In 1879 Gately, being hounded by a group of 'young roughs', sought the protection of the police who locked him up for being drunk, prompting Gately to try to cut his own throat with a knife. It took three police to prevent his suicide. Hangman Jones succeeded where Gately failed, killing himself to avoid the notoriety of hanging a woman, Frances Knorr (see Chapter 13). And Elijah Upjohn, Ned Kelly's hangman, was finally forced to flee Melbourne in fear of his safety.

Governor Castieau, of the Melbourne Gaol, showed little enthusiasm for witnessing hangings and floggings and described himself as 'stupidly sensitive & after my long training inconsistently anxious about my duty' (Castieau's Diaries, 11 June 1877, Finnane 2004). He would often retreat to his club for a 'nobbler', or a drink or two, after such grisly occasions and it may be possible that the nature of his work contributed to his failing health.

The difficulties of his position had a more tragic outcome for a later governor, Mr William Clark, who shot himself in the grounds of the Gaol in 1921. It was reported in the press of the day that he 'had worried a great deal when three prisoners, charged with capital offences, were given into his custody'.

Fiona Leahy and Chris Briggs

In the hanging of the bushrangers Andrew George Scott (aka Captain Moonlite) and Thomas Rogan in Sydney in 1880, Scott was reported to have 'died instantaneously' but Rogan 'struggled convulsively for a few minutes', even though the prison doctor 'considered Rogan was insensible from the time he fell' (*Wagga Wagga Advertiser* 1880).

The pages of the Newgate calendar (recording the judicial executions in London) are replete with bungles, to the point that many could perhaps have been more accurately described as attempted executions! Not infrequently the rope broke – sometimes twice, as in the case of the pirate Captain Kidd in 1701 (Crook 1926) – or the noose slipped from the condemned man's neck, as in the case of William Snow in 1789 (James and Nasmyth-Jones 1992). The reverse can also occur, with evidence that, in some cases, the drop has been such that it resulted in decapitation.

In colonial Australia there were several examples of hangings that did not go well. The *Argus* newspaper of 8 January 1887, seven years after Ned Kelly was hanged, reported the bungled hanging of four prisoners convicted of gang rape:

19th-century post mortem dissection practices and the Pentridge remains

Post mortem procedures in 19th-century Melbourne may have followed the methods of Rokitansky (Vienna, 1804–1878) or Virchow (Berlin, 1821–1902), most likely the latter. Following external examination of the body, a horizontal incision is made across the front of the chest with a second incision downwards towards the pubis. The ribs are cut and the anterior chest wall removed to expose the underlying thoracic viscera. Virchow's technique involved sequential dissection, separation and examination of each organ with 'painstaking observation and application of scientific principles' (Hill and Anderson 1988).

An incision over the top of the head from ear to ear was followed first by folding of the scalp forward, then by sawing of the cranium horizontally at a point just above the orbital margins and 1 cm above the most prominent part of the skull, posteriorly. A hammer and chisel might be required to assist removing the skull cap. The brain is then exposed with or without its covering dura membrane.

Detailed descriptions of hangings in the Sheriff's Particulars of Execution indicated that in the majority of cases some form of post mortem examination took place. Early reports on the executions of Martha Needle and Ernest Knox (both in 1894) have no specific reference to autopsy. Subsequent reports have greater detail, for example in relation to Frederick Jordan: 'Dr Shields making a most minute post mortem ...'

There is also physical evidence of autopsy having taken place. Saw marks are evident on some of the intact skulls, and are even visible on fragments of skulls as well as on cervical vertebrae. These marks indicate an examination of the brain or cervical spine, the latter being to infer the success of the hanging – cervical fracture or dislocation being closely correlated with the rapidity of death.

Unusually, in all cases but one, where we found evidence of fracture or dislocation of the vertebrae there are no saw marks. It is possible that where fracture or dislocation was determined by examination of the neck, and the medical officer was satisfied that death was instantaneous, no further dissection took place. An English gaol surgeon, Dr William Caute, described his dissection technique as follows:

> *In making an incision along the vertical spines the finder ran suddenly into a cavity between two of the bones, which were discovered to be the third and fourth cervical vertebrae and there was also a gap of nearly an inch between these vertebrae.*

Certainly the two female skulls from the Pentridge remains, likely to be Knorr, Needle or Williams, are intact with no signs of dissection, even though the Particulars of Execution stated that each woman had a post mortem examination and that in Needle's case 'an autopsy was made', confirming that the neck was broken or dislocated.

References

Hill RB, Anderson RE (1988) *The Autopsy: Medical Practice and Public Policy*. Butterworth, Boston.

The Particulars of Executions 1894–1967, Sheriff's Office of Victoria. Series VPRS 14526.

Fiona Leahy and Chris Briggs

> The ropes were quickly adjusted, and at a sign from the hangman the bolt was drawn, and the four prisoners were suddenly left struggling in the air. Then followed a most horrible scene. The drop was at least a foot too short, and instead of the prisoners being killed outright by the breaking of their necks, as should have been the case, they were slowly strangled to death, the struggles of Boyce being terrible to witness. It was bad enough to see Read drawing deep breaths for what seemed fully a couple of minutes after the drop had taken place, but the contortions made by Boyce lasted far longer, and his clenched fists showed how hard had been the final struggle. Martin in falling was caught by the arm in a bight of the rope, and hung for half a minute with his pinioned elbow bent up under his ear. After this lapse of time his arm was jerked loose by an agonised spasm, and he dropped about a foot. The whole affair appeared to have been terribly bungled, and the sight was altogether too sickening to describe with anything like fullness.

Public executions were banned in England in 1868 as the public nature of the punishment was deemed not to be essential to its alleged deterrent effect (and although the last judicial execution was in 1964 the death penalty was not removed until 1998). In Australia, the last public execution was that of Francis Thomas Green, who was convicted of murder and hanged outside Darlinghurst Gaol in Sydney on 21 September 1852, and the last judicial execution was that of Ronald Ryan in 1967 (see p. 116). Public executions still occur in several countries: Amnesty International has stated that Iran, North Korea, Saudi Arabia and Somalia have all conducted public executions in recent years (*Guardian* 2013).

The scale of drops

In 1886, a parliamentary committee on capital sentences in England (Marshall and de Zouche 1888) proposed a scale of drop distances to create an impact force to the neck of 1260 feet pounds (or 5600 N). So, a man weighing 154 pounds (70 kg), would require a drop of 8.2 feet (2.5 m) to achieve the desired force of 1260 foot pounds. There is no clear basis for this number, though. The committee also appeared to prefer a subaural (i.e. below the ear) position for the knot of the noose, rather than an occipital (back of the head) or submental (beneath the chin) position.

Certainly a knot placed under the chin runs the risk of not resulting in serious injury to the cervical spine and, because the leading edge of the noose itself is at the back of the neck, if the cervical spine is not sufficiently damaged to result in spinal cord damage, the front of the neck will not be compressed, and the hanging will fail (James and Nasmyth-Jones 1992).

Reference to a knot is probably relevant for the 19th century. Certainly in more recent times, the knot has been replaced by a metal eye through which the noose is created. The cowboy coil-type hangman's noose seen in the movies is an American phenomenon.

Modern examinations of hanging

In 1993, a 31-year-old man was judicially hanged in Washington state in the USA and a 7/8 inch braided hemp rope was used. The drop was 7 feet (2.1 m) and the the knot consisted of six loops. The victim's head was covered with a black felt hood and the noose placed outside this, with the knot immediately below the left ear. No heart sounds were heard or peripheral pulses felt one minute following release of the trapdoor.

Post mortem CT and MRI imaging and an autopsy were performed on the body, revealing:

> ...disruptions of anterior and posterior longitudinal ligaments with disc herniations at (the level of the second and third and fifth and sixth cervical vertebrae), and small non-displaced fractures of the right transverse process of C5 and the superior end plate and left transverse process of C6. Spinal cord imaging was normal (Reay *et al.* 1994).

In addition, there was a midline fracture of the thyroid cartilage (the adam's apple).

Actual post mortem dissection showed as well:

- extensive haemorrhage in the muscle and soft tissue of the neck;
- fractured left wing of the hyoid bone (a small horseshoe shaped bone at the base of the tongue);
- bleeding (subarachnoid) over the base of the brain.

The common expectation is that judicial hanging will produce the 'hangman's fracture', which is a fracture of the second and third cervical vertebrae, a consequence of which is bruising or tearing of the upper spinal cord, resulting in death. Such damage was not demonstrated in this case, although a fatal outcome had occurred in under one minute.

A US study of the skeletal remains of 34 judicial hanging deaths between 1882 and 1945 found very few actually had cervical fractures. The opportunity for such a study had arisen as exhumations were required to move the remains of prisoners from prison grounds as prisons were decommissioned and sold to developers – as happened with Pentridge Prison in Melbourne. The study found there were only seven fractures in total, six of which were of the axis (the second cervical vertebra).

The execution of Ned Kelly

Ned Kelly's drop from the gallows at the Old Melbourne Gaol (see Fig. 11.1) on 11 November 1880 was described by the *Argus*:

> The noose having been adjusted, the white cap was pulled over his face, and the hangman step-ping to the side quickly drew the bolt, and the wretched man had ceased to live. He had a drop of 8ft., and hung suspended about 4ft. from the basement floor. His neck was dislocated and death was instantaneous; for although muscular twitching continued for a few minutes, he never made a struggle. It was all over by five minutes past 10 o'clock, and was one of the most expeditious executions ever performed in the Melbourne gaol (*Argus*, 12 November 1880).

LAST SCENE OF THE KELLY DRAMA: THE CRIMINAL ON THE SCAFFOLD.

Fig. 11.1: Last scene of the Kelly drama: the criminal on the scaffold. *Australasian Sketcher*, 20 November 1880. State Library of Victoria.

References

Argus, 12 November 1880.
Argus, 8 January 1887.
Crook GT (1926) *The Complete Newgate Calendar*, Vol. 2. Navarre Society, London.
Finnane M (ed.) (2004) *The Difficulties of My Position: The Diaries of Prison Governor John Buckley Castieau, 1855–1884*. National Library of Australia, Canberra.
Guardian, 12 April 2013.

James R, Nasmyth-Jones R (1992) The occurrence of cervical fractures in victims of judicial hanging. *Forensic Science International* April, **54**(1), 81–91.
Laurence J (1926) *A History of Capital Punishment,* Sampson, Low, Marston and Co., London.
Marshall JJDZ, de Zouche JJ (1888) Judicial execution. *British Medical Journal,* October.
Pierrepoint A (1977) *Executioner: Pierrepoint.* Hodder and Stoughton, Great Britain.
Reay DT, Cohen W, Ames S (1994) Injuries produced by judicial hanging. *American Journal of Forensic Medicine and Pathology* **15**(3), 183–186. doi:10.1097/00000433-199409000-00001.
Wagga Wagga Advertiser, 24 January 1880.

Chapter 12
The prison governor

Mark Finnane

The prison governor who oversaw the execution of Ned Kelly was an interesting character, and an analysis of his diaries reveals much not only about the man and his attitude to the prisoners under his charge, but about the times he lived in.

I was working on the history of prisons in Australia in the late 1980s when I came across the extraordinary diaries of John Buckley Castieau, one-time governor of Old Melbourne Gaol (see Fig. 12.1). The Castieau diaries were donated by family descendants to the National Library of Australia about 30 years ago. Their range is impressive. In 12 volumes, dating from the first entries in 1855, during the state trials of the Eureka rebels, to the last, in 1884 only a month before Castieau's death, they provide an outstanding record of daily life in colonial Victoria during those years. Early in 2013 a single volume, the 1875 diary, was recovered from a deceased estate – it has now been added to the NLA collection. Castieau was a dreamer, a drinker, an aspirant writer, a player in amateur theatre, a conversationalist, a Victorian husband and father of six children. From his base at the Melbourne Gaol he issued out daily to visit the Melbourne clubs, attend the theatre and stroll the city streets with his wife and children. In the evenings, in spite of his fondness for a 'toddy', he also proved a diarist of some discipline, over long periods of time. As rich as the diaries are, the surviving volumes have a glaring gap – the period during which Ned Kelly was in Melbourne Gaol, first on trial, then in the death cells, and finally executed on a day long remembered, 11 November 1880. It is tempting to speculate that a diary might exist for that momentous year in which Kelly became the most famous inmate of Old Melbourne Gaol. Yet the erratic record of the single foolscap volume that includes entries from 1877–79 suggests Castieau had by that time wearied of the discipline required by his previous practice of daily entries.

What we do have in the many earlier volumes is the best available account of what it was like to be a prison governor, managing a prison, overseeing its discipline, supervising floggings and witnessing executions, all the while keeping the contemporary media informed of the latest news about the prison. The closest we get to what might have been Castieau's interactions with Ned Kelly in his last days might be in the diary record of May 1872 relating to another hanging:

> In the evening did not go out as I had a man under Sentence of Death who was to be executed to-morrow. After nine o'clock I went over to see this man who was Edward Feeney charged & convicted of having murdered Charles Marks … Feeney was particularly quiet & when I spoke to him said he was quite

Fig. 12.1: Examination and remand of Ned Kelly in Melbourne Gaol. *Australasian Sketcher*, 14 August 1880. Governor Castieau is seated across the table from Ned Kelly, with the large moustache. State Library of Victoria.

> prepared to meet his fate. He however wished he said to make a statement & that was to deny the allegation that he had been told had been published of his having been improperly intimate with the murdered man Marks. This he solemnly declared to be untrue & as a dying man declared there was not the slightest grounds for the rumour. Feeney wished me to make this statement of his public & I agreed to do so. Did not go out this evening but was interviewed by a Reporter of the *Argus* to whom I told my tale. He of course was delighted to get it as it was something approaching the Sensational (Finnane 2004).

The next day Castieau recorded the chilling detail of a badly performed hanging (the prison doctor 'could not however answer my question as to why some when executed were troubled with reflex movements & others were not') and the subsequent forensic inquiries as the doctors attempted to establish whether the executed man had a 'vicious indulgence' (Kaladelfos 2013). In such episodes we can learn of the business of managing corporal punishment and execution from the viewpoint of a servant of the Crown, who was a humanitarian in disposition and sentiment, and a witness to the death of Ned Kelly.

The archival record details many of the governor's dealings with Ned in his last days, including transcriptions of Kelly's last letters that attempt to explain and excuse his actions (Jones 1995). There are many references to Castieau's interactions with Ned Kelly in the press of the day, with newspapers such as the *South Australian Register* reporting on 10 July 1880, after Ned Kelly was admitted to the Melbourne Gaol and was in the prison hospital recovering from his wounds:

> As he appears to be stronger, the prisoner was permitted by the medical officer to see his mother, whom he had urgently requested to see. Mr. Castieau, the Governor of the gaol, was present during the interview, which lasted for a considerable time. Mr. Castieau cautioned Kelly to keep calm, and he replied that he would, requesting Mr. Castieau to give the same advice to Mrs. Kelly. The mother seemed to feel acutely pained by the intelligence of the affray at Glenrowan, and at seeing the condition of her son. Kelly was in a tolerably communicative mood and conversed freely with his mother.

Fact or fiction: A film about Ned Kelly was the world's first feature film

It is often reported that Charles Tait's 1906 film, *The Story of the Kelly Gang*, was the world's first full-length feature film. The previous contender for that title was the Salvation Army's *Soldiers of the Cross*, made in 1900, but research revealed that was actually an elaborate mix of slides, live acting and only about 13 minutes of film.

The Story of the Kelly Gang, of which only about a quarter of the original hour-long film remains, cost around £1000 to make and over the years of its screening returned 25 times that amount. Its first screening was at the Athenaeum Hall on 26 December 1906. It was not without controversy, though, and was seen not just to glorify crime but to inspire it. According to the National Film and Sound Archives, in May 1907 a screening in Ballarat prompted five local children to break into a photographic studio to steal money, then bail up a group of schoolchildren at gunpoint. This resulted in the Victorian Chief Secretary banning screenings of the film in Benalla and Wangaratta, both towns with strong Kelly connections.

For many years *The Story of the Kelly Gang* was thought lost, but segments were found in various locations and collections, including some found on a rubbish dump. In 2007 the film was inscribed on the UNESCO *Memory of the World Register* for being the world's first full-length feature film.

Other films about or based on Ned Kelly and the gang have included *The Kelly Gang* (1920), *When the Kellys Were Out* (1923), *When the Kellys Rode* (1934), *A Message to Kelly* (1947), *The Glenrowan Affair* (1951), *Stringybark Massacre* (1967), *Ned Kelly* (1970), *Reckless Kelly* (1993), *Ned Kelly* (2003), *Ned* (2003) and a telemovie *The Last Outlaw* (1980).

An interesting note is that the son of the Governor of Old Melbourne Gaol, Godfrey Castieau, is said to have met the outlaw when he was 13 years old, and later played Ned Kelly under the stage name Godfrey Cass in the silent films *The Kelly Gang* and *When the Kellys Were Out.*

Craig Cormick

In November of that year, Castieau had the task of supervising Kelly's execution, as reported in the *West Australian* of 23 November 1880:

> Mr. Castieau, the governor of the gaol, informed the condemned man that the hour of his execution was fixed for 10 o'clock. Kelly simply replied, 'Such is life.' His leg irons were then struck off, and after a short time he was marched, accompanied by several warders, from the condemned cell in the old wing of the prison to the central building. He was very submissive, and on the way, passing through a portion of the gaol grounds, laid out in flower-beds, he remarked, 'What a nice little garden.' He said nothing further until reaching the pressroom, where he remained until the arrival of Father Donahy, the chaplain of the gaol.

References

Finnane M (2004) *The Difficulties of My Position:The Diaries of Prison Governor John Buckley Castieau 1855–1884.* National Library of Australia, Canberra.

Jones I (1995) *Ned Kelly: A Short Life*. Lothian Books, Melbourne.

Kaladelfos A (2013) 'Until death does part us': male friendship, intimacy, and violence in late colonial Australia. In *Intimacy, Violence and Activism: Gay and Lesbian Perspectives VII.* (Eds G Willett and Y Smaal) pp. 39–55. Monash University Press, Melbourne.

Ned Kelly in gaol. *South Australian Register,*10 July 1880.

The execution of Ned Kelly. *West Australian,* 23 November 1880.

Chapter 13
Who were the other prisoners executed and buried at the Melbourne Gaol?

Fiona Leahy and Chris Briggs

To better understand the context of Ned Kelly's execution and burial at the Old Melbourne Gaol, it is useful to examine the other prisoners who were executed and buried there, and whose remains the Victorian Institute of Forensic Medicine analysed. They represent not just a cross-section of criminals who were sentenced to death in the late 19th century, but their stories reveal insights into the crimes and times of 'Underbelly' Melbourne of that era.

Ned Kelly was one of 135 prisoners executed at the Melbourne Gaol between 1842 and 1924, and just one of at least 34 prisoners buried within the Old Labour Yard of the gaol (see Chapter 4). Those other prisoners laid alongside him included three women and people of many races.

Martha Needle

On 22 October 1894, 14 years after Ned Kelly's death, Martha Needle walked 'boldly to the scaffold, and remained firm to the last in her resolve to make no confession' (*Clarence and Richmond Examiner* 1894). She showed no remorse for the slow and painful deaths of her husband, children and a lodger, who she poisoned with 'Rough on Rats' (*Advertiser* 1894) – clearly just as rough on humans. Her motive was never truly established. It was suggested that she poisoned her lodger, Louis Juncken, because he opposed her marriage to his brother, Otto. It is more likely that Needle was insane.

However, this defence was rejected by the trial judge, Justice Hodges, who stated that the law 'cannot wait while philosophers, physicians, and scientists settle metaphysical problems' regarding insanity (Cannon 1994). The officials at the Melbourne Gaol admitted that they were at a loss to understand this woman who had paid scant attention to the ministrations of the clergy and appeared self-righteous to the end. They regarded her as 'inscrutable as the sphinx' (Cannon 1994).

Frances Knorr: the 'baby farmer'

Just nine months earlier, on 15 January 1894, Frances Knorr, the 'baby farmer', was hanged for the murder of two babies placed in her care by desperate and impoverished women (see Fig. 13.1). In the terrible depression of the 1890s, orphanages and charitable institutions in Melbourne were filled to overflowing with abandoned children and destitute women. 'Baby farming' became established as a private alternative, where women unable to look after their children paid the baby farmer, usually another woman, a small weekly fee to feed and house their child. The first attempt to regulate this practice was introduced by the Victorian

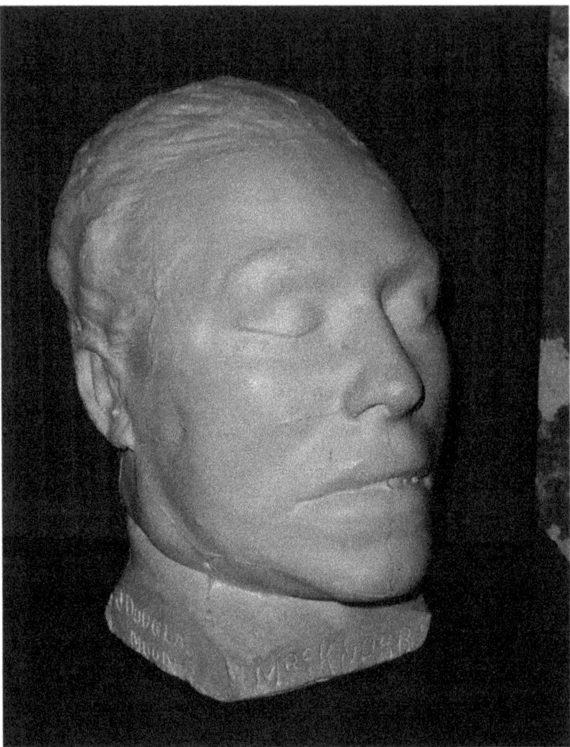

Fig. 13.1: The death mask of Frances Knorr at the Old Melbourne Gaol. Ronn Taylor.

government in the *Public Health Amendment Act 1883* which required baby farmers to register their house with the local Board of Health. Many women, including Frances Knorr, evaded these regulations (Harris 2010).

The execution of Frances Knorr elicited a very different response from those both inside and outside the Melbourne Gaol than following the death of Martha Needle. On the eve of her execution she had admitted to the murder of two babies in a written confession to the Premier, Mr Patterson, and showed deep remorse for their deaths. Frances Knorr was a struggling young mother herself and the scene in the gaol when she surrendered her young child was a very moving one: 'the woman clung with desperation to her babe, and great scalding tears streamed down her cheeks and showered on the innocent face of the baby girl as she bade her good-bye' (*Queensland Times* 1894).

Frances Knorr was the first woman to be executed at the Melbourne Gaol since Elizabeth Scott in 1863. This provoked a public debate on whether women should be subjected to capital punishment. In a pyrrhic victory for equality of the sexes, the governor advised that women should not be spared the gallows on the basis of their sex. The prospect of executing a woman proved too much for hangman 'Jones', however, who 'hacked his throat with a razor' and bled to death in his gaol room just one week before Knorr's scheduled execution (*Advertiser* 1894).

In the minutes leading up to her death, a crowd had gathered outside the Melbourne Gaol extending down the entire block to Swanston St. Many were women 'who relieved their overcharged feelings with tears, though nothing of course could be seen' (*Dubbo Liberal and Macquarie Advocate* 1894). The silence in the gaol was 'broken by the sound of voices attuned to a song of praise, softly at first and then more clearly' as Reverend Mr Scott, the female warders, prisoners and Frances herself sang the hymn 'Safe in the arms of Jesus' followed by 'Abide with me' (*Queensland Times* 1894). Their voices faded as Frances Knorr was led to the scaffold. Her final words were reported to be, 'I do not care what man can do to me. I have peace, perfect peace' (*Dubbo Liberal and Macquarie Advocate* 1894).

Emma Williams

The plight of Emma Williams, executed the following year on 4 November 1895, is equally troubling. She drowned her small son in the Port Melbourne lagoon, weighing down his body with a large stone held in place with a piece of braid torn from her dress. The papers declared that she had abandoned her husband 'for a gay life'. But you have to wonder – Williams was married at 14 to a man twice her age, became a mother by 15 and was a widow at 20 after her husband died in 1893 from typhoid fever.

Williams was left penniless with a child to support and little means of earning money (Cannon 1994). She had tried to 'palm the child off' on various people so that she could go to work, at one time earning money as a tailoress (*South Australian Chronicle* 1895). She had also attempted to obtain admission for the child to a charitable institution, but entry was refused for both mother and child (*Bendigo Advertiser* 1895).

In sentencing Williams, Justice Hodge showed little pity. He believed that 'she desired to be relieved of the child so that she might be able to repeat further acts of immorality' (*Bendigo Advertiser* 1895). Williams had been forced to support herself through prostitution.

She sought a stay of execution on the grounds that she was 'quick with child' (pregnant) (*Chronicle* 1895). However, upon examination during the trial Dr Shields advised that this was not so, and he confirmed that diagnosis in post mortem examination (*Argus* 1895). Williams was deeply penitent at the gallows and in a written message even forgave the principal witness who testified against her (*Riverine Herald* 1895).

Perhaps Williams was unlucky in being arrested and charged for this terrible crime. In the late 19th century infanticide was seen to be a great social problem in the eastern colonies. It was reported in press that:

> … during the last three years in the Melbourne Morgue alone enquiries have been made concerning ninety-nine cases of infant deaths, and sixty-two of them were pronounced to have been the result of wilful murder. Last year one instance of deliberate extinction of child life was recorded on the morgue books on the average of every 12 days, and yet the conviction or even prosecution of the responsible criminals is very rare (*South Australian Register* 1894).

These three women were among the 34 or more prisoners buried in the Old Labour Yard within the prison grounds, along with Ned Kelly, after the enactment of the *Criminal Law and Practice Statute 1864*.

Multicultural profile of executed prisoners

The profile of the executed prisoners buried at the Old Melbourne Gaol is illustrative of the multicultural nature of the Victorian colony at that time. Although most of those buried at the gaol were originally from England (17) and Ireland (11), others were from Switzerland, Russia, China, India, America, Italy and the Philippines. Given that many of the executions took place during the era of the Victorian gold rush it is easy to conclude that most of the murders were for theft on the gold fields; however, this seems to have rarely been the case.

One instance from the gold fields was the 'treacherous' James Seery, who was executed for murdering a digger and his dog, while stealing 'parcels of gold and some money' at Good Luck Creek in Victoria. A Chinese man, An Gaa, was sentenced to death for killing 'a fellow countryman named Pooey Waugh' over a claim on the banks of the Loddon River (*Argus* 1875). The majority of the remainder were for murder, often in the act of robbery, or crimes domestic in nature. Arthur Buck, Patrick Smith, Henry Howard, Arthur Oldring, Frederick Jordan and William Colston, for instance, were all convicted for murdering their wives (or lovers) following quarrels.

Identifying the remains of Ronald Ryan

In February 1967 Ronald Ryan, aged 42, was executed by hanging, having been found guilty for the shooting and subsequent death of prison officer George Hodson, during an escape from Pentridge Prison two years earlier. Ryan is famous as the last prisoner executed in Australia.

As it was known that his body was buried in an unmarked grave at Pentridge Prison, attempts were made to identify his remains in order to release them to his family.

One coffin with no name plate was located and found to contain the remains of a middle-aged adult male of Caucasoid ancestry with a stature of 166–179 cm (see Chapter 4). Interestingly, the coffin, which was buried on its own, was turned north–south with the head placed at the south end, unlike the traditional coffin placement of east–west with the head at the west. Also of interest, the coffin contained a thick white substance covering the body inside, most likely lime that was traditionally used to hasten decomposition. However, there are two types of lime and in some cases the addition of lime can actually preserve remains.

The white casing was removed to reveal the relatively complete remains of an adult skeleton containing all the upper and lower dentures. There were impressions of woven fabric, presumably remnants of clothing, on the undersurface of the casing. A small amount of human hair was also preserved.

Identification through DNA analysis was attempted but it was not possible to extract either nuclear or mitochondrial DNA from the skeletal remains.

Subsequently, craniofacial superimposition was attempted. A photograph of the skull of the deceased was laid over and compared with a photograph of Ryan when he was alive. The results of the superimposition showed no inconsistencies between the skull and the photograph. Taking into account the circumstances of burial, together with the results of the forensic anthropology and superimposition analyses, the skeletal remains were formally identified as Ronald Ryan and in early 2008 released to the family for proper burial.

Soren Blau

Who was Frederick Deeming?

There will never be any debates about whether Frederick Bailey Deeming was a hero or a criminal. He was about as bad as they come. He was a swindler, bigamist, thief, seducer and multiple murderer.

Frederick Deeming. State Library of Victoria.

He murdered his first wife and four children in England in 1891 then fled to Australia, where he remarried. He murdered his second wife later the same year.

He had first moved to Australia in 1882 at the age of 29 with his first wife, Marie, returning to England with considerable wealth in 1889, believed to have been obtained from a diamond swindle in South Africa that Deeming took part in. In England he married a young woman bigamously, before deserting her and fleeing to South America.

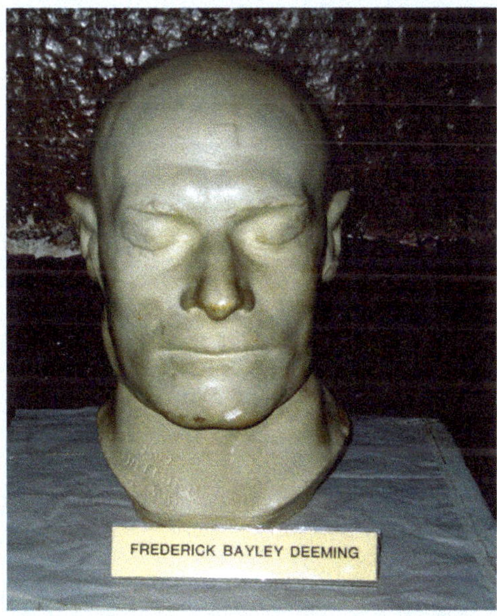

Death mask of Frederick Deeming at the Old Melbourne Gaol. Ronn Taylor.

He then married another woman, Emily Mather, and moved to Australia with her in 1891. They arrived in Melbourne on 15 December and 10 days later he viciously murdered her, most likely cutting her throat and bashing her head in. He buried her under the hearthstone of a parlour. The body was found after a later tenant complained about the 'disagreeable' smell coming from the bedroom. Deeming had long fled and the press and public were aghast at the savagery of the murder.

Using the names Williams and Barron Swanston, Deeming began looking for a new wife, but the police were on his trail and caught up with him in Fremantle. At the same time the murder of his first wife in England was discovered, with her body and that of their four children found buried under the kitchen of their house in Rainhill, near Liverpool. The children were Bertha (nine), Marie (seven), Sidney (five) and Leala (18 months).

The brutal savagery of the murders and Deeming's apparent lack of remorse made his case a national and international sensation, with many newspapers trying to link him with Jack the Ripper.

Despite a plea of insanity he was found guilty and hanged on 23 May 1892.

An autobiography which Deeming wrote in gaol, perhaps giving some insights into his mind, was destroyed by the authorities.

Craig Cormick

Future identifications

Aside from Ned Kelly, only two other prisoners have been identified and their remains released to family descendants for private burial: Colin Ross and Ronald Ryan. Ryan was the last person to be executed in Victoria, in 1967 (see p. 116), and Colin Ross was executed for the murder of Nell Alma Tirtschke in 1922 – he was posthumously pardoned for the crime.

To date the VIFM has not identified any other individual prisoner from the group of remains exhumed from Pentridge Prison in 2008 and 2009, even though the Institute has received several inquiries from descendants of executed prisoners seeking information about the possibility of identification of their forebears. Unfortunately none of the families could locate a suitable maternal descendant to enable mitochondrial DNA analysis and comparison to be made. And even if we were able to obtain suitable family reference samples for individual prisoners, mtDNA was not able to be obtained from all the remains.

References

Argus, 22 July 1875.

Cannon M (1994) *The Woman as Murderer*. Todays Australia Publishing Co., Mornington.

Execution of Emma Williams. *Argus*, 5 November 1895.

Execution of Emma Williams. The child slayer dies calmly. *Riverine Herald*, 5 November 1895.

Execution of Frances Knorr. *Dubbo Liberal and Macquarie Advocate*, 17 January 1894.

Execution of Martha Needle, an inscrutable psychological problem. *Clarence and Richmond Examiner*, 23 October 1894.

Harris DH (2010) Victoria Police involvement in the *Infant Life Protection Act 1893–1908*. *Journal of Public Record Office Victoria* 9.

Martha Needle's career. *Advertiser*, 23 October 1894.

The condemned woman Williams. *Chronicle*, 26 October 1895.

The execution of Frances Knorr. *South Australian Register*, 15 January 1894.
The execution of Frances Knorr. *Queensland Times, Ipswich Herald & General Advertiser*, 8 February 1894.
The Melbourne Hangman Jones cuts his throat with a razor. a remnant of fine feeling. He would not hang Mrs Knorr. *Adelaide Advertiser*, 8 January 1894.
The Port Melbourne child murder. *South Australian Chronicle*, 24 August 1895.
The Sandridge child murder. *Bendigo Advertiser*, 25 September 1895.

Chapter 14
Reading Ned's head: colonial phrenology, popular science and entertainment

Dean Wilson

Even though the practice of phrenology was declining and falling into disrepute by the time Ned Kelly was executed, his head was still subjected to a phrenological reading and his character analysed from it, which was done in the name of science.

On the morning of 11 November 1880, Ned Kelly's body fell from the 8 foot (2.5 m) high scaffold in Melbourne Gaol. Within minutes, the notorious outlaw swung lifeless at the end of the hangman's noose. But the colonial authorities were not finished with him just yet. After a short time, probably less than an hour, Ned's body was pulled down, piled into a cart and taken to the prison dead house. There the beard and hair of his corpse were shaved, and plaster applied to the still-warm skin of his face and head in order to make a death mask (Jones 1995) (see Fig. 14.1). Twenty-four hours later a wax rendering of Ned Kelly's head was on display in Bourke Street (*Herald* 1880).

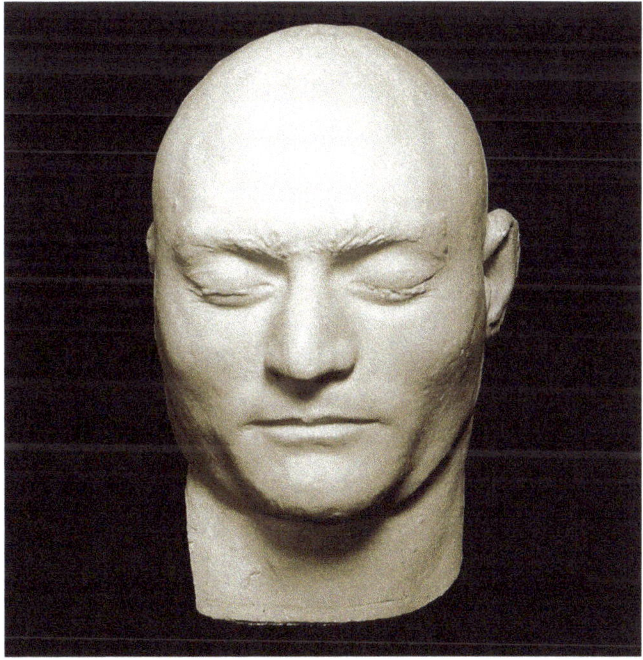

Fig. 14.1: Ned Kelly's death mask. National Trust Australia (Victoria).

Ned Kelly was already famous – or infamous, depending upon your perspective – in the colony of Victoria and further afield. Consequently Kelly's death mask would have provoked an unusually high degree of public curiosity. Nevertheless, the taking of death masks from executed criminals was common practice in the 19th century, and the seemingly bizarre and macabre sequence of events following Kelly's death were no more than the routine procedure of prison authorities. Taking casts of the heads of freshly executed criminals was a well established practice in Victoria by the mid 1850s, and this mirrored similar developments in England (Gatrell 1994).

In 1855 the *Argus* newspaper reported that casts of the heads of James Condon, John Dixon and Alfred Jackson, three men recently executed for highway robbery, had been made by one M. Pardoe (*Argus* 1855). In the 19th century death masks featured in waxworks museums and the collections of prison authorities and as instructive exhibits for popular 'scientific' lectures on crime and criminality.

A significant number of these 19th-century death masks have survived and several Australian institutions hold significant collections. The Melbourne Gaol retains a sizeable collection of death masks of executed criminals taken by Victorian prison authorities, while the New South Wales Justice and Police Museum also has a significant collection. The State Library of Victoria has a Ned Kelly death mask (the exact number of these in existence remains unknown). It also has another important mask in its collections – that of notorious murderer Frederick Bailey Deeming (see Figs 14.2, Fig. 14.3).

Fig. 14.2: Frederick Deeming's death mask (side view) at the Old Melbourne Gaol. Ronn Taylor.

Fig. 14.3: Frederick Deeming's death mask (view from above) at the Old Melbourne Gaol. Ronn Taylor.

Why were death masks made?

The answer to this question is actually quite complex, as 19th-century death masks had several interwoven purposes and meanings. They were concurrently objects of colonial power, objects of entertainment and objects of science. Their meaning as objects of colonial power remains evident to us, and is not so difficult to decipher. On one level, the taking of death masks from criminals provided a material display of colonial authorities' victory over crime, and particularly of their victory over such ominous threats to colonial order as bushranging. What clearer way could there have been to express the unassailable power of colonial authority than assembling a collection of transgressors' heads?

The fascinated stares of contemporary museum visitors encountering death masks suggests that the magnetism of these wax likenesses as objects of entertainment endures into our own times. In the 19th century crime and criminals were no less the subjects of popular fascination than they continue to be today. In Sydney in 1844, a cast was made of the head of John Knatchbull, a convict who bludgeoned a woman to death in a shop, and a woodcut later produced from the cast was offered for sale to the public (Kociumbas 1992). Figures of criminals were also a staple display of Melbourne's waxworks (Colligan 1987) Murderers, unsurprisingly, were always of interest but colonial Australians were particularly preoccupied with bushrangers.

Such was the depth of this fascination that some even attempted to obtain body parts of dead outlaws as curios. When bushranger Johnny Gilbert's corpse remained at the Binalong police station for several days in 1865, numerous locks of hair were stolen from the body before burial (Penzig 1983) A popular rumour of the period maintained that medical men had cut off the scrotum of Dan 'Mad Dog' Morgan the bushranger, killed by police in 1865, which some rather ghoulish individual then took to using as a tobacco pouch (Carnegie 1977). The colonial enthralment with criminals pervaded all levels of colonial society: John Buckley Castieau, Governor of Melbourne Gaol, is said to have run a small commercial enterprise supplying death masks to acquaintances (see Chapter 12).

While it is still comparatively easy for us to comprehend death masks as tokens of colonial power and the props of popular entertainment, it requires a greater leap of historical imagination to understand their 19th-century relevance as scientific objects. Nevertheless, in the 19th century death masks were taken primarily to further scientific inquiry, not to amuse or to provide evidence of the authorities' victory over crime.

Death masks, once systematically subjected to scientific analysis, were believed to offer a key to interpreting and understanding the character of the criminal, as a material manifestation of the widespread Victorian faith in physiognomy – the prevalent belief that individual character could be read through the facial features and the forms of the head and body (Cowling 1998). These ideas lay behind the emergence of a new science in the early 19th century that was to have considerable impact in Europe, America and Australia – the science of phrenology.

Phrenology

The founding figure of phrenology was Viennese physician Franz Joseph Gall (1758–1828), who carried out research in prisons and asylums on the 'diseased brains' of criminals and the insane. Drawing from this research, he hypothesised that the human brain contained almost 30 different departments. Gall suggested that these 'faculties', combined with the nervous system, were the driving force of all human activity. The character of every individual was thus suggested to be formed from the myriad potential combinations of these 'faculties' and it was the relative size of each, he suggested, that determined individual character. Significantly, phrenology supposed a close enough correspondence between the outer surface of the skull and the shape of the brain to enable the proficient observer to recognise the relative importance of the faculties (Cooter 1984; Davies 1955; de Giustino 1975; McLaren 1974). Consequently, phrenology at a popular level was the art of reading character from 'bumps' on the head. English satirists mockingly dubbed the new science 'bumpology' (McLaren 1981).

Phrenology enjoyed great popularity in the Australian colonies from the 1830s and, while adherents to the tenets of the science were primarily drawn from the upper working class, several significant colonial administrators of the late 1830s and early 1840s were zealous phrenologists (Kociumbas 1992). Prison reformer Alexander Maconochie, who was appointed Superintendent of the Norfolk Island penal settlement in 1840, was closely connected to phrenological circles in Britain, and gave lectures on phrenology to his fellow passengers on his voyage out to the colony (de Giustino 1972). Asylum reformer Dr Francis Campell, Superintendent of Tarban Creek Lunatic Asylum in Sydney from 1848, was also an ardent phrenologist (Roe 1965).

The enthusiasm for phrenology was initially centred in Sydney, but the influence of craniological theories soon spread to Melbourne. Lectures on phrenology were given soon after Melbourne was established as a settlement, with the *Port Phillip Herald* advertising a public lecture to be given on the subject as early as 1840. The standing of phrenology was also boosted by some powerful devotees among Victoria's colonial elite, including Archibald Michie, who had lectured on the subject at the Sydney School of Arts before his arrival in Melbourne in 1852 (Nadel 1957).

Popular lectures on phrenology attracted significant crowds keen to witness an array of self-designated 'Professors' extrapolate on the links between head shape and character. Producing the head of a notorious criminal for illustrative purposes guaranteed a lecture would be standing room only. Criminals were a preferred topic for phrenologists, providing as they did a means of interpreting a supposedly flawed character. However, it is important to note that phrenology was viewed as a reforming science. The characteristics identified by the phrenologist were, once understood, amenable to transformation and the criminal could be trained in the ways of righteousness (de Giustino 1975). Nevertheless, phrenologists more often lectured on criminals whose chances of reformation had been brusquely ended by their execution.

Melbourne possessed its own experts in the field and in the 1850s Philemon Sohier, a self-proclaimed 'Professor of Phrenology', had phrenological heads prepared by a professional wax modeller (Colligan1987). In 1855, Sohier delivered a phrenological lecture using the death masks of the earlier-mentioned highway robbers James Condon, John Dixon and Alfred Jackson as props. Sohier's lecture was delivered to a capacity crowd at the Temperance Hall in Russell St, and the *Argus* noted that full houses could be anticipated for any future lectures using the casts (*Argus* 1855).

The credibility of phrenology as a serious science gained significant momentum in Victoria when the colonial government commissioned and published *A Phrenological Report on Aborigines*, prepared by Philemon Sohier, as part of the proceedings of the Select Committee on Aborigines in 1858. Phrenology retained a strong popular appeal into the 1860s, largely as it was considered to be a science with practical application that could be easily learnt by the dedicated layperson. This rendered phrenology strongly attractive to a lower middle-class instilled with an ideology of self-education and self-improvement. Practitioners of phrenological science such as Dr Blair, Dr William Edward Crook and J.W. Ffrost continued to deliver lectures to enthusiastic audiences at Mechanics Institutes and town halls throughout the 1860s (*Age* 1860).

The grotesque and extraordinary

Phrenology lectures were a heady mix of popular science and entertainment. Phrenologists offered their audiences scientific education, while their focus on criminals simultaneously satiated a popular appetite for the grotesque and extraordinary. Gradually, however, much of phrenology's early scientific veneer washed away, and it progressively descended into the realm of popular amusements.

In Britain the scientific credibility of phrenology had been diminishing since the 1840s (de Giustino 1975). Phrenology's status as serious science lasted longer in colonial Victoria, however, where it retained serious supporters into the 1860s. Nevertheless, by the end of the decade, the appetite for instructive lectures on the topic had dissipated. By the mid 1870s advertisements for phrenology lectures in Mechanics Institutes had vanished from the pages of Melbourne's newspapers, as the phrenologists migrated into the city's amusement arcades, setting up shop next to the stalls of the palm readers and fortune-tellers.

Ned Kelly's death mask

By the time Ned Kelly's death mask was taken in 1880 phrenology had largely disappeared as a subject of discussion in serious scientific circles. Formerly published in the finest learned journals of the Empire, phrenology was now ridiculed as a pseudoscience – the preserve of quacks, charlatans and eccentrics. Nevertheless phrenological ideas continued to percolate through Melbourne's colonial culture at a more popular level.

One phrenologist, Mr Hume, based in the Victoria Arcade, claimed to be expert at 'pointing out the most suitable occupations' of his customers and informing them of 'the natural bent of their minds'(Phrenologist Hume). Another phrenologist, Professor Shepherd, offered a like service at the Eastern Market, where for two shillings and sixpence individual customers were informed 'what you are best adapted for' and were given their own chart as part of the price of a head reading (*Age* 1880). An example of a phrenological chart issued by one Eastern Market phrenologist to patrons is held in the Monash University Rare Books Collection; it includes an engraving of Ned Kelly to illustrate overdeveloped 'animal propensities' (Brown 1883). Popularisers of the 1880s, such as Shepherd and Hume, provided phrenology for mass consumption – still vaguely self-improving, but with a heavy emphasis on amusement and diversion.

Given phrenology's status by the 1880s, it is perhaps unsurprising that the cast for Ned Kelly's death mask was taken by a man who straddled the worlds of science and entertainment. The cast was taken by waxworks proprietor Maximilian Ludwig Kreitmayer (see p. 64). Kreitmayer was a skilled medical modeller who had studied anatomy in Munich and Glasgow, and went into partnership with Melbourne's resident phrenological 'Professor', Philemon Soheir, in 1863. His skills were regularly engaged by the Victorian government, which had called upon him to produce displays for Imperial Exhibitions in London and at the Paris Exhibition of 1878. His 'Aboriginal Groups' were also in demand for official exhibitions (Colligan 1987). Kreitmayer had an established record of government contracts and was an obvious choice to take a cast of Ned Kelly's head.

A death mask of the notorious bushranger was inevitably going to stir enormous public curiosity. One man in particular, A.S. Hamilton, was to surface as the expert on Kelly's head and its interpretation. Adopting the moniker 'Professor of Phrenology', Hamilton was a 'travelling phrenologist' who had given lecture tours in the Australian and New Zealand colonies since the 1860s. He claimed to have been a 'practicing phrenologist' since the 1840s (*Herald* 1880). While it would be wise to treat many of Hamilton's claims with some scepticism, there is some truth in his assertion that he was a practising phrenologist.

He was the author of a pamphlet on the subject of phrenology and had performed phrenological examinations of prisoners awaiting execution (and made casts of their heads afterwards) in New South Wales and New Zealand before arriving in Victoria (Hamilton 1866). Ever the showman, Hamilton clearly sensed considerable commercial potential in the possibility of reading Kelly's skull. Wasting no time, Hamilton wrote to the Chief Secretary on 10 November 1880 asking permission to make a phrenological examination of Kelly before his execution. This would, he claimed, have the virtue of 'throwing the light of science upon the character of the condemned man'. Hamilton also offered to make a cast of Kelly's head for the government.

It is not surprising that Hamilton's request was denied. Hamilton as much as conceded that his request was unlikely to be successful, ending his correspondence with a note that he hoped the Chief Secretary would 'make an exception in my favour under present very peculiar circumstances' (VPRS 1880). The 'peculiar circumstances' to which Hamilton referred were not only the general passions surrounding Kelly's execution. Hamilton was President of the Society for the Abolition of Capital Punishment, an organisation which had been a thorn in the side of the government since the sentence of death was pronounced on Kelly (Brown 1948).

Hamilton's objection to capital punishment was derived from that of the English phrenologists, who argued that executions badly affected the minds of spectators and that the criminal should be treated to rectify cranial abnormalities rather than being dropped from the scaffold (de Giustino 1975).

Measuring Kelly's head

Hamilton did not get to see Ned Kelly before his execution, but there is evidence that he was there while the cast for the death mask was taken. Max Brown's study, *Australian Son*, stated that he was not only present as Kreitmayer took the cast but even assisted by taking measurements of Kelly's head (Brown 1948).

Several days later Hamilton used the Kreitmayer cast as the basis for a phrenological study that was published in the Melbourne *Herald* on 18 November 1880. Apart from using the opportunity as a platform for gratuitous self-promotion (Hamilton suggested a phrenological examination of all prisoners held in Melbourne and Pentridge gaols should be undertaken – by himself, of course), Hamilton assured readers he would shed the light of science, 'the custodian and revealer of truth', upon the character of Kelly. Hamilton's study suggested Kelly had several dangerously overdeveloped cranial regions, including those of combativeness, destructiveness and love of approbation. Underdeveloped in the outlaw were the qualities of cautiousness, sublimity and conscientiousness. Hamilton's lengthy phrenological analysis concluded that most of Kelly's law-breaking could be ascribed to his monstrously overdeveloped self-esteem. Hamilton cautioned readers that heads the shape of Kelly's threatened the very fabric of society:

> ...there are few heads amongst the worst that would risk so much for the love of power as is evinced in the head of Kelly from his enormous self-esteem. This self-esteem, combined with large love of approbation combined with hope, would often make him make appear bright, dazzling and heroic to those who could not see through the veil which vanity threw around him (*Herald* 1880).

Was the phrenology of Ned Kelly taken seriously?

The Ned Kelly death mask provided a powerful symbol of the colonial authorities' ultimate victory over the outlaw, and was undoubtedly an object of great public curiosity. But would anyone have really taken Hamilton's phrenological analysis seriously? The answer is more than likely yes. As historian Roger Cooter suggested, despite losing serious scientific credibility 'phrenology in the second half of the century became in many ways more deeply entrenched than ever in everyday thought and expression' (Cooter 1984).

Victoria's intellectual elite may have scoffed at Hamilton's phrenological analysis, which essentially recycled popular images of Kelly to warn of the dangers of elevating him to heroic status. But for many readers, Hamilton's analysis – with its numbers, categories and measurements – would have projected the image (if not the substance) of learned scholarship.

Ned Kelly's head attracted a grotesque fascination but it also remained, for many, an object of science.

References

A phrenological report on Aborigines for the Victorian Parliament. Report of the Select Committee on Aborigines 1858. *Votes and Proceedings of the Legislative Council*, session 1858–59.

Age 14 May 1860, 3 September 1860, 13 November 1860.

Age 30 May 1861, 26 September 1861.

Argus 3 December 1855.

Argus 30 November 1885.

Brown M (1948) *Australian Son: The Story of Ned Kelly*. Georgian House, Melbourne.

Brown R (1883) *A Delineation of the Character, Talents, Physiological Developments and Natural Adaptations of Mr.————————*. Mason, Firth & McCutcheon, Melbourne.

Carnegie M (1977) The death of Morgan the bushranger. *Victorian Historical Magazine* 2(188), May.

Colligan M (1987) Canvas and wax: images of information in Australian panoramas and waxworks with particular reference to Melbourne 1849–1920. PhD thesis, Monash University.

Cooter R (1984) *The Cultural Meaning of Popular Science: Phrenology and the Organization of Consent in Nineteenth-century Britain*. Cambridge University Press, Cambridge.

Cowling M (1998) *The Artist as Anthropologist: The Representation of Type and Character in Victorian Art*. Cambridge University Press, Cambridge.

Davies JP (1955) *Phrenology, Fad or Science: A Nineteenth-century American Crusade*. Yale University Press, New Haven.

de Giustino D (1975) *Conquest of Mind: Phrenology and Victorian Social Thought*. Croom Helm, London.

de Giustino D (1972) Reforming the commonwealth of thieves: British phrenologists and Australia. *Victorian Studies* **15**(4), 439–461.

Gatrell VAC (1994) *The Hanging Tree: Execution and the English People 1770–1868*. Oxford University Press, Oxford.

Hamilton AS (1866) *Practical Phrenology: A Lecture on the Heads, Casts of the Heads, and Characters of the Maungatapu Murderers, Levy, Sullivan, and Burgess*. Examiner Office, Nelson.

Herald 18 November 1880, 12 November 1880.

Jones I (1995) *Ned Kelly: A Short Life*. Lothian Books, Melbourne.

Kociumbas J (1992) *Oxford History of Australia. Vol. 2: Possessions 1770–1860*. Oxford University Press, Oxford.

McLaren A (1974) Phrenology: medium and message. *Journal of Modern History* **46**, 86–97.

McLaren A (1981) A prehistory of the social sciences: phrenology in France. *Comparative Studies in Society and History* **23**, 3–22. doi:10.1017/S001041750000966X.

Nadel G (1957) *Australia's Colonial Culture: Ideas, Men and Institutions in Mid-nineteenth Century Eastern Australia.* Harvard University Press, Cambridge, Mass.

Penzig EF (1983) *A Real Flash Cove: The Story of the Bushranger John Gilbert.* Kangaroo Press, Kenthurst.

Phrenologist Hume, 15 Victoria Arcade Melbourne. Pamphlet, State Library of Victoria collection.

Port Phillip Herald 1 September1840.

Roe M (1965) *The Quest for Authority in Eastern Australia 1835–1851.* Melbourne University Press, Melbourne.

Victoria Public Record Series 6888. Part V, Kelly Historical Collection. A.S. Hamilton, 21 Collins Street East to Chief Secretary, 10 November 1880.

Chapter 15
The science of the Kelly gang's armour: distilling fact from fiction

Gordon James Thorogood

Studying the composition of the metal of one of the Kelly gang's armour allowed researchers to determine what temperature it had been heated to. This provided an answer to the long-debated question of whether the armour had been made by a blacksmith or on a bush forge. As a charcoal fire and bellows used by a blacksmith can heat metal to over 1000°C, but a bush forge could only get the metal to cherry red, or about 750°C, the scientific distinction can be made.

Nothing defines Ned Kelly and the Kelly gang more than their suits of armour. But as an iconic image the armour has attracted many myths, often perpetuated though song, stories, paintings, silent features and movies, which have blurred the lines between fact and fiction.

It is known, for instance, that the Kelly gang manufactured the famous suits of armour in the period between the murders of the three police in the Wombat Ranges in October 1878 and the siege at Glenrowan in June 1880. But how exactly they were manufactured, and by whom, is less certain.

It has been asserted by some prominent historians that the Kelly gang had an accomplice in a willing blacksmith who fabricated the armour from ploughshares and that, in fact, several people could have been involved in the fabrication of the armour. This suggests that the gang had some support from their fellow settlers, who may have also been having a difficult time with the authorities.

Another theory, however, is that the gang forced the local blacksmiths to supply materials and fabricate the suits. But given that word did not get out in the time between the construction of the suits and the siege at Glenrowan, this seems unlikely. The third alternative is that the bushrangers made the suits themselves, from whatever materials they could source, be they ploughshares or not, using a simple bush forge in a remote location. This possibility would indicate that the bushrangers had a fair amount of skill at working metal, and fits in well with the version of the legend that imagines them to be a group of passionate, intelligent men at odds with the establishment of the time.

It is not possible to overstate the interest and passion aroused by the Kelly legend in Australia, hence the debate about the fabrication of the armour is important to historians and the public alike.

But how to test for the truth? Looking beyond the stories and the myths, the answer came from the metal in the suits of armour. And the story of the analyses undertaken of the metal of the armour provide some insight into the method of how the armour was constructed.

Fact or fiction: The Kelly gang's armour was inspired by traditional Chinese armour

There are two leading theories for the inspiration for the armour worn by the gang. The first is that was based on the types of Chinese armour that members of the gang had seen worn during a carnival procession through the streets of Beechworth in 1873. Joe Byrne had a close relationship with some Chinese people in the area and was reported to have been able to speak some Cantonese.

The second key theory is that Ned got the idea from his favourite book, *Lorna Doone: A Romance of Exmoor*, by English author Richard Doddridge Blackmore (1869). The story is about an outlaw family living in a valley in Exmoor in south-west England. In one part, describing the outlaws, it states:

> *the horsemen passed in silence, scarcely deigning to look round. Heavy men and large of stature, reckless how they bore their guns, or how they sate their horses, with leathern jerkins, and long boots, and iron plates on breast and head, plunder heaped behind their saddles, and flagons slung in front of them; I counted more than thirty pass, like clouds upon red sunset.*

Four suits of armour are known to have been manufactured and there are persistent stories that other sets were also made for some of their sympathisers, but no evidence of them has ever emerged. Brian Cookson, a journalist who travelled through Kelly country to research the Kelly gang for a series of newspaper articles published in the Sydney *Sun* in 1911, wrote of the difficulty of tracking down exactly who had manufactured the armour: 'During the course of a few casual inquiries, we found that the armour was made, solely and individually, by each of eleven blacksmiths, living in all parts of the territory.'

Some of the Kelly gang's armour collected after the siege at Glenrowan. State Library of Victoria.

There was some mixing up of the gang's armour after it was recovered from the burned inn at Glenrowan, and the National Trust recently sought to correctly identify the sets. Joe Byrne's armour was believed complete, but the other suits of armour were found to be quite mixed up with a mix of helmets and breast plates. The proper suits were not reassembled until 2003.

The custodians of the suits are:

- Steve Hart's armour is held by the Old Melbourne Gaol (National Trust);
- Dan Kelly's armour is held by the Police Historical Unit (Police Museum in Melbourne);
- Joe Byrne's armour is in private hands;
- Ned Kelly's armour is held by the State Library of Victoria.

Craig Cormick

The testing was only possible courtesy of Rupert Hammond, who is the owner of Joe Byrne's suit of armour. He is the descendant of Sir William Clarke, who was given the armour by Police Superintendent Hare, who claimed it after it was taken from Joe Byrne's dead body at Glenrowan. Mr Hammond made it available for scientific analysis just before its inclusion in a special exhibition on outlaws held at the National Museum of Australia in 2003.

How was the armour tested?

A team at the Australian Nuclear Science and Technology Organisation's (ANSTO) facility at Lucas Heights, NSW, used a combination of several scientific techniques to analyse the armour (Figs 15.1, 15.2). These included:

- neutron and X-ray diffraction;
- gamma-X-ray fluorescence;
- optical metallography;
- transmission electron microscopy.

These techniques are described in more detail in this article, as well as how the results of these investigations were able to indicate how the armour was made.

The particular techniques chosen were quite important as they had to be relatively non-destructive, considering the suits of armour are regarded by many as national treasures and as such should not be damaged in any way in the analysis process. However, as discussed later, the owner did allow some destructive testing on a very limited area of the armour.

To allow for a more thorough analysis of the armour it was disassembled into its component parts – helmet faceplate, helmet cylinder, breast plate, back plate, lap plate, two side panels and the rivets (see Fig. 15.3).

Because the armour was made of steel and over 100 years old, some surface oxidation, or rust, had occurred. Not knowing how deep this may have been, and not wishing to disturb the surface, it was decided to use neutron diffraction to analyse each metal piece of the armour. Neutrons are subatomic particles and neutron radiation can penetrate most materials

Fig. 15.1: The author (walking backwards) delivering the armour to ANSTO amid much media attention. ANSTO.

Fig. 15.2: The researchers with the armour. ANSTO.

How was the helmet worn?

Anyone who has ever tried on a replica of one of the Kelly gang's helmets was immediately struck by how heavy they are and how much they cut into the collar bone, and probably wondered how it was possible to wear them for prolonged periods.

The answer is that the weight of the helmet was not meant to be borne on the collar bones. A close examination of the helmets shows there are two holes punched at the top on the front, back and sides of each helmet. These were used for stringing leather straps through which most of the helmet's weight was borne on the top of the wearer's head.

Ned Kelly is reported to have been wearing a woollen cap at Glenrowan, to pad his head from both the helmet and the straps.

Craig Cormick

The police were warned about the armour

One of the police informants, using the code word 'diseased stock' to refer to the gang, wrote to Assistant Commissioner Nicolson, warning the police of the manufacture of the armour as early as May 1880. The letter, using coded terms, warned that stolen ploughshares had been turned into armour and had been tested to withstand gun fire at 10 yards. The letter stated:

> *Nothing definite re the diseased stock of this locality. I have made careful inspection, but did find [sic] exact source of disease. I have seen and spoke to and on Tuesday, who were fencing near home. All others I have not been able to see. Missing portions of cultivators described as jackets are now being worked, and fit splendidly. Tested previous to using, and proof at 10 yards. I shall be in Wangaratta on Monday, before when I may learn how to treat the disease. I am perfectly satisfied that it is where last indicated, but in what region I can't discover. A break out may be anticipated, as feed is getting very scarce. Five are now bad. I will post a note giving any bad symptoms I may perceive from Wangaratta on Monday or Tuesday at latest, and will wait on you for news how to proceed on a day which I shall then state, before end of the week. Other, animals are, I fear, diseased. -Yours faithfully, B.C.W.*

However, Superintendent Hare, who later took over from Nicolson, stated at the Royal Commission into the Kelly Outbreak in 1881 that he heard about the armour from another source:

> *When I relieved Mr. Nicolson, on the 2nd of June 1880, he gave me no information whatever concerning the Kelly armour, and it was only in conversation with different people – Mr. Sadleir especially – that I picked up that, and found that it was supposed that when the Kellys next appeared it would be in armour.*

Craig Cormick

very easily – they will pass readily through a steel plate. Some of the neutrons reflect off the atoms that are arranged in rows or more specifically planes to produce peaks in a neutron diffraction pattern. This technique is similar to X-ray diffraction. Both diffraction methods make use of Bragg's law, a scientific discovery made by William Lawrence Bragg and his father William Henry Bragg only 32 years after the siege at Glenrowan. Australia's first Nobel prize, awarded in 1915, recognised that discovery.

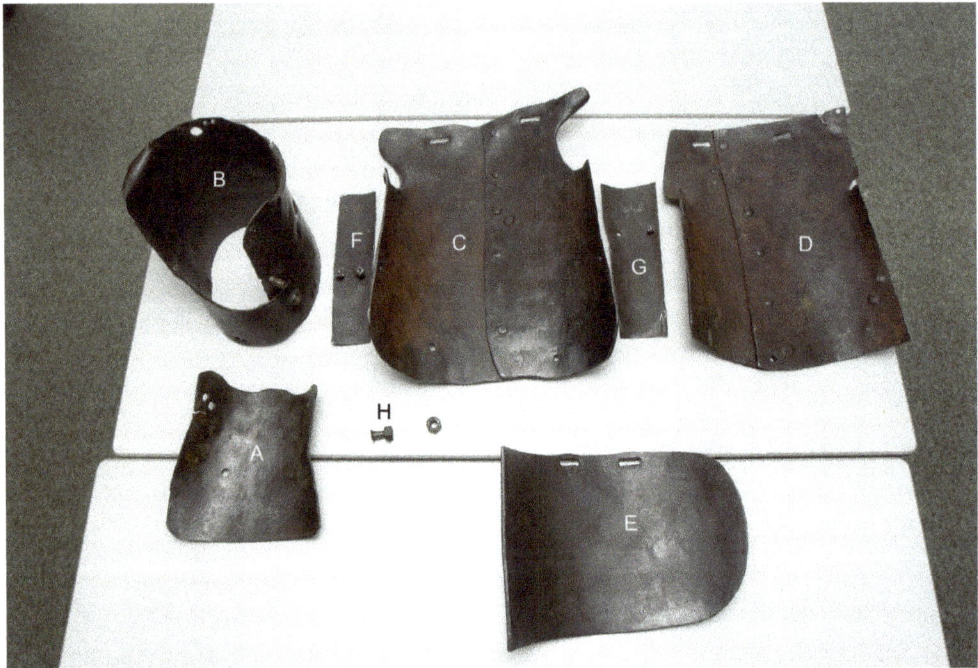

Fig. 15.3: Component parts of armour. (a) Helmet face plate. (b) Helmet cylinder. (c) Breast plate. (d) Back plate. (e) Lap plate. (f) Side panel 1. (g) Side panel 2. (H) Rivet. ANSTO.

ANSTO has a triple axis diffractometer, which was used to collect neutron diffraction patterns (see Fig. 15.4). Comparison of the diffraction patterns of the different pieces showed that they were essentially the same metal (see Figs 15.5, Fig. 15.6). The patterns also indicated that the metal was steel with a strong 'preferred orientation' (meaning there was a tendency for metal crystals to be oriented more one way than another), as would have occurred had the steel been rolled.

There were minor differences in preferred orientation in each of the steel pieces examined, but they were sufficiently similar for us to conclude that the different pieces came from similar source material (Creagh *et al.* 2004).

X-ray diffraction

The next technique used on the armour was X-ray diffraction. Unlike neutrons, X-rays penetrate only a few microns into the surface of most materials, the depth dependent on the material being examined. Compared with neutron diffraction, X-ray diffraction can be considered as a technique which samples the surface of the object rather than deep inside it. X-ray diffraction patterns were collected on a device known as an X-ray diffractometer, which was fitted with an energy-discriminating detector to make it particularly sensitive.

We would have normally required the surface being examined to be flat, but reconfigured the diffractometer to account for the curved shapes of much of the armour. The one drawback to the X-ray diffraction was the limited size of objects that could be examined by it, and so only the smaller side plates were examined using this technique.

Fig. 15.4: Helmet mounted for neutron diffraction. ANSTO.

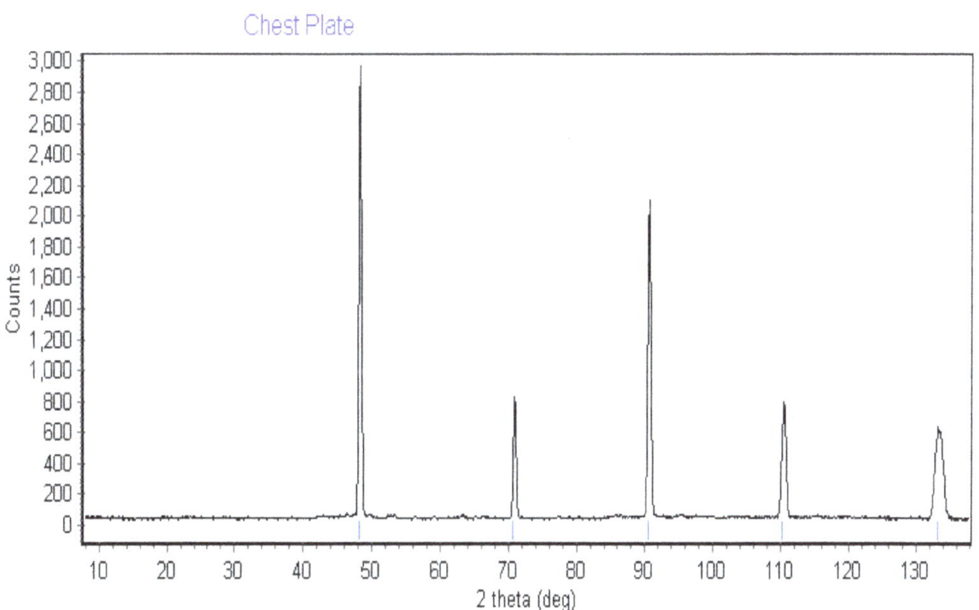

Fig. 15.5: Neutron diffraction pattern of breast plate. ANSTO.

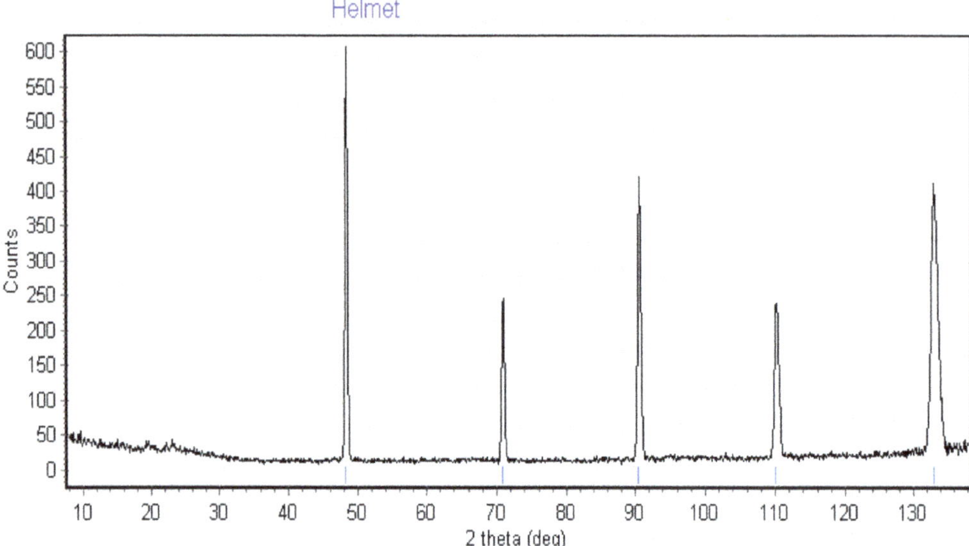

Fig. 15.6: Neutron diffraction pattern of helmet. Note intensity variation in the last two peaks compared to Fig. 15.2 due to preferred orientation. ANSTO.

The results, shown in the peaks and troughs (see Fig. 15.7), showed they were very typical of steel and also indicated a surface that had been highly stressed. This would indicate that the side plates had been cold worked (beaten without being heated), which was not consistent with the typical practice of blacksmithing at the time.

Gamma-X-ray fluorescence spectroscopy

The next test undertaken was using gamma ray emitting radioisotopes, which can 'excite' the electron shells of the various atoms within a metal. When the electrons drop back to their

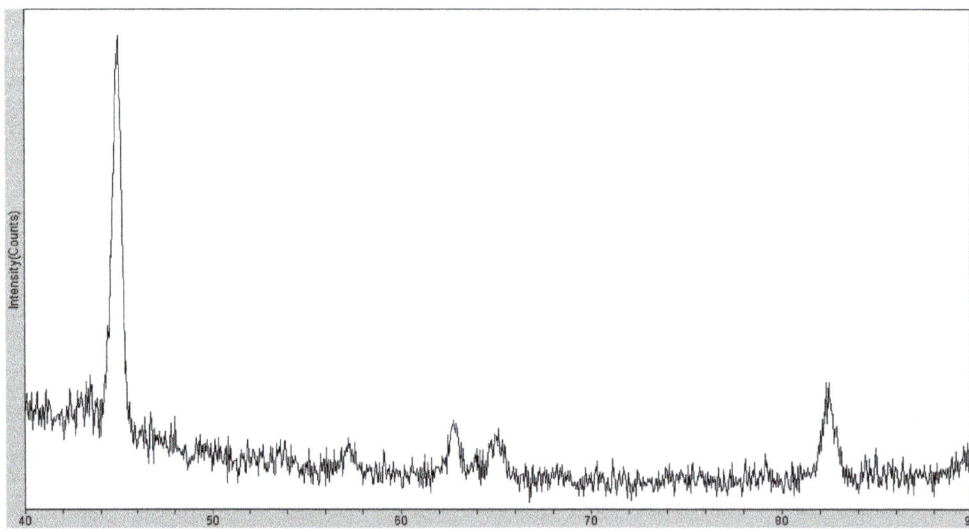

Fig. 15.7: X-ray diffraction of side plate. ANSTO.

Table 15.1: Weight percentage compositions of the major atomic species in samples of Joe Byrne's armour, examined by gamma X-ray fluorescence spectroscopy.

Location		Manganese (%)	Iron (%)	Arsenic (%)	Tin (%)	Lead (%)
Helmet	Face plate	0.32	96.56	0.17	1.80	0.18
Helmet	Cylinder		94.6	0.1	2.6	0.18
Breast plate	General	0.33	96.85	0.07	1.48	0.14
Back plate	General	0.32	96.56	0.17	1.80	0.18
Lap plate	General	0.31	94.29	0.1	2.6	0.18
Side panel 1	General		95.33	0.11	2.6	0.18
Side panel 2	General		98.14	0.25	0.47	0.09
Rivet	Typical		97.65		1.34	
Carbon steel		0.3	96.0			

Note that this technique cannot determine carbon content. The comment 'general' means that the reading is the average reading taken from various locations on the piece.

normal positions afterwards, they emit energy in the form of X-rays. Each atom gives off a characteristic X-ray energy which can be considered the 'signature' for that atom. This allows us to measure the various energies emitted by the atoms in the metal, then identify and assign percentages of each element present. And this tells us the exact amounts of different elements in a piece of metal, which in turn can be used to help identify its location and date of manufacture.

Table 15.1 shows the compositions of most of the components of the armour, which were all rather similar. The major element present was iron, at between 94% and 96% with minor amounts of manganese, arsenic, tin and lead. Of note, some of the pieces of armour, notably the cylindrical helmet, side panels and rivets, did not contain any traces of manganese. From this we can deduce that they were likely taken from different source materials from the rest of the steel pieces. Certainly the side panels were much less thick than the rest of the armour, and the rivets were made from steel rod that was cut and banged flat on each end. It is quite possible that the material for the helmet came from a different, or older, stock than the rest of the armour.

Optical metallography

It had now been established that the components of the armour were made from carbon steel, but the techniques used did not enable us to draw conclusions about the carbon content nor the metallurgical state of the steel, and whether all parts of the steel had the same metallurgical structure. Optical metallography had to be used to answer these questions. This, however, required the destruction of some of the metal, to which Rupert Hammond consented.

Much scientific analysis rests on developing suitable tests, then developing processes to conduct the tests. To a layperson this may look like putting samples into a machine, pushing a button and getting the required result, but most such machines are simply sophisticated tools to assist in conducting different analytical experiments. In the case of optical metallography, the first stage of the experiment is to get a sample of a compound from within the metal known as carbides (these are composed of carbon with another element) within the metal. This is done by etching the surface with the acid nital, to dissolve the surface layers of metal, leaving the carbides standing away from the surface (they are not dissolved by nital). A softened acetate sheet is then placed on this surface and peeled off once it has hardened. Several carbides are then transferred from the surface of the metal onto the sheet, showing their position as they would have been in the metal and thus replicating the original microstructure.

The acetate replica is then examined by standard optical microscopy, as a representative sample of the surface. It should be mentioned that very large carbides, where the long axis of the carbide is not parallel to the etched surface, may not be sufficiently revealed by etching to be released onto the acetate sheet. Such failed extractions can be identified, though, as they manifest as holes or mounds in the replica, indicating regions where large carbides were present in the original microstructure.

Optical metallography has the huge advantage that it can reveal whether the steel had been above its plastic transformation temperature (Samuels 1992). This is the temperature at which metal bends easily rather than having to be forced, and thus shows the strains of such force upon it. If the steel had not been taken above this temperature, it was very unlikely that a blacksmith had been involved in its manufacture.

The first part of the armour examined was the back plate (see Fig. 15.3). An area on the armour had a wear mark, from which most of the surface material had already been removed. This provided an ideal opportunity to examine the bulk microstructure of the steel. Examination of the replica showed fine grains of ferrite 70% (an iron oxide component) and pearlite 30% (a combination of ferrite and cementite, forming distinct bands or layers in any slowly cooled carbon steel) (see Fig. 15.8). This ratio indicates a carbon content of approximately 0.2%, which is consistent with mild steel.

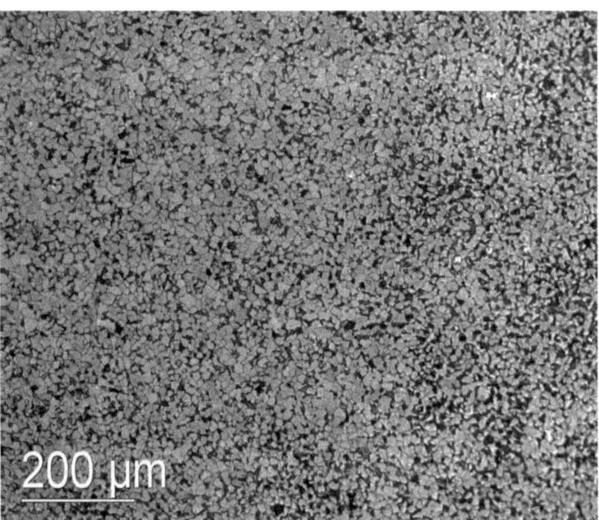

Fig. 15.8: Taken from the back plate, lower inside surface. Note the fine equiaxed grains of ferrite (light 70%) and pearlite (dark, 30%) ASTM grain size 8. Note the swirl pattern of pearlite concentration. ANSTO.

Fig. 15.9: Coarse carbide plates in the pearlite phase. ANSTO.

The microstructure was consistent with correctly produced rolled steel of the day. When large slabs of metal are melted down and cast for rolling purposes, alternating layers of manganese and silicon occur. This segregation causes pearlite to form in the high manganese areas and ferrite to form in the high silicon areas, which in turn produces banding of pearlite and ferrite in the original material, as found in the samples tested.

We also knew that if the steel had been worked above 723°C there would have been some evidence in the shape and structure of the pearlite present (see Figs 15.9, 15.10). But analysis indicated that the metal had not been heated to a very high temperature, estimated at only 500°C and 700°C.

Transmission electron microscopy

Turning to the next process for examining the armour, a single acetate replica was taken from the lower corner of one plate in the centre of the breast plate. This replica represented the plate viewed edge-on. The replica was prepared in the same manner as those prepared for optical microscopy, but the acetate replica was coated with a very very thin layer of carbon (120 nm). The acetate was then dissolved in acetone to leave the carbon extraction replica floating free, the carbides being held in place by the very very thin carbon film. This carbon extraction replica was then examined by transmission electron microscopy, a way of measuring the atomic content of the extracted particles by illuminating them with a focused beam of electrons.

The purpose was to look for any non-carbide materials, such as sulphides, which might differentiate the steel from more modern steels, which tend to be free of sulphur contamination. Analysis had also indicated high levels of tin and lead in this steel.

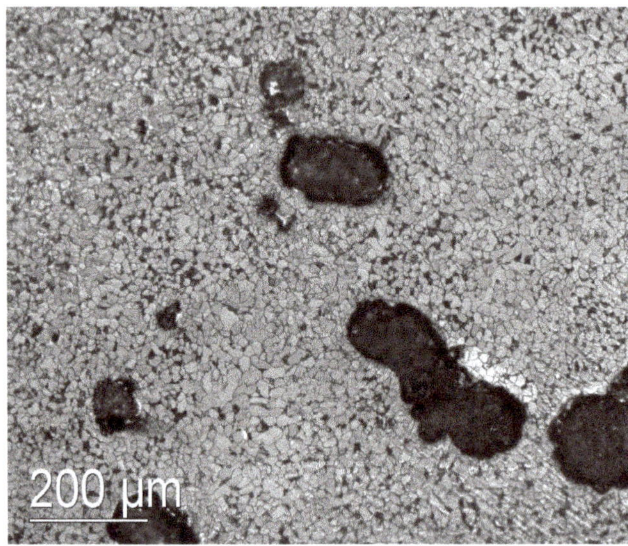

Fig. 15.10: Pearlite at edges of holes and lack of deformed grains indicates metal through loss by corrosion. ANSTO.

Analysis of the extracted carbides was achieved by focusing an electron probe onto the area of interest and analysing the X-rays emitted. Regions as small as 10 nm can be readily analysed – to give context of size, the width of a human hair is around 80 000 nm.

How mild steel is produced

The results showed the characteristics of mild steel – a little background on the production of steel will be helpful to understand its composition. The first affordable industrial-scale process for producing large amounts of steel from molten pig iron was developed by Henry Bessemer and patented by him in 1855. The Bessemer process for producing steel involved blowing air through molten pig iron to remove impurities and reduce the carbon content to produce steels of various compositions.

The next major advancement was the open hearth process, licensed by French engineer Pierre-Emile Martin in 1865, whereby oxidation of the carbon and other impurities was achieved by exposing a shallow bath of molten pig iron to an oxidising atmosphere, and by the addition of oxidants such as iron ore and lime.

Both these processes produce steel that contains dissolved oxygen, which bubbles and froths out of solution when the steel is cast, producing a spongy metal with poor properties. Robert Mushet solved this problem in 1857 by de-oxidising molten steel with manganese. The manganese combined with oxygen in the melt to form manganese oxides, which separate to a slag (off-products of the steel production). Manganese also has the very beneficial property of combining with undesirable sulphur in the steel to form globules of manganese sulphide. This locks up the sulphur and prevents the formation of grain boundary films of iron sulphide, which would cause the steel to crumble when hot worked. Therefore, mild steels invariably contain residual manganese from the de-oxidising and de-sulphurising process, as well as manganese sulphide inclusions – which were present in the metal of the armour.

Although analysis of the steel by gamma X-ray fluorescence spectroscopy (see Table 15.1) indicated unusually high tin and lead concentrations, these elements were not found using transmission electron microscopy. Overall the microstructure appeared entirely consistent with mild steel which had been air cooled from above its A1 temperature (the temperature at which the steel undergoes a change that can be measured). In the case of mild steel, this temperature range is about between 650 and 723°C. Examinations indicated that this steel had not been cold worked, but had been subjected to some heat.

Conclusion

What did it all mean? Neutron diffraction, X-ray diffraction and gamma X-ray fluorescence spectroscopy all showed that the steel used in Joe Byrne's armour was a high-grade rolled carbon steel, similar in composition and structure to that used in ploughshares of the period. The presence of manganese in most components of the armour also confirmed that the steel in those pieces was produced after 1858. It is possible that the steel in the helmet, side plates and rivets came from steel objects from an earlier period.

Optical metallography determined that pearlite was present in its original form, as would have been evident in the steel at the time of manufacture. Other areas displayed evidence of the steel being heated only to a dull red colour. We can therefore conclude that:

- the armour was made by amateurs rather than blacksmiths, possibly by the gang;
- it was made at low temperatures such as would be attained in an open bush forge;
- it was shaped using available round objects such as tree trunks;
- the different pieces of metal came, for the most part, from a common origin;
- the chemical composition was consistent with high-quality rolled carbon steel that would have been used to manufacture ploughshares.

Visual examination of the armour (see Fig. 15.11) confirmed that the major pieces seemed to be from ploughshares, as the boltholes in it occurred at a spacing consistent with those in contemporaneous ploughshares and there were actually maker's markings on some pieces. It has not been possible to determine the makers, however.

It should be noted that after this body of work was concluded the author was contacted by a person claiming to have some of the parent material used in the manufacture of the armour. Subsequent examination of this material via the same techniques led to the conclusion that it was not of the same material as that used to construct the armour of Joe Byrne. This does, however, highlight the fact that a thorough examination of the other suits has not been performed. To fully conclude if all the suits were made of the same material by the same methods as Joe Byrne's armour, such an examination does need to be performed.

And perhaps we should end on a reminder that Joe Byrne died from a gunshot wound to the groin while sitting at the bar drinking – despite the effort that went into its manufacture, his armour failed to protect him.

Fig. 15.11: Joe Byrne's armour assembled. ANSTO.

Acknowledgements

The research underpinning this chapter was undertaken by G.J. Thorogood, D.C. Creagh, M. James, G.K. Smith, D.L. Hallam, D.R.G. Mitchell, G. Atkinson and R. Hammond.

References

Creagh DC, Thorogood GJ, James M, Hallam DL (2004) Diffraction and fluorescence studies of bushranger armour. *Radiation Physics and Chemistry* **71**(3), 839–840. doi:10.1016/j.radphyschem.2004.04.107.

JCPDS (2004) *PDF4 Set of Relational Databases.* International Commission for Diffraction Data. Newton Square, Pennsylvania.

Jenkins R, de Vries J (1967) *Practical X-Ray Spectrometry.* Philips, Eindhoven, The Netherlands.

http://www.statelibrary.vic.gov.au/slv/exhibitions/kellyculture/armour.html.

Hunter BA (1998) Rietica: a visual Rietveld program. *Commission on Powder Diffraction Newsletter* **20**, 21.

Parrish W, Wilson AJC, Langford JI (1998) X-ray diffraction methods: polycrystalline. In *International Tables for Crystallography,* Vol. C. 2nd edn. Section 5.2, pp. 487–497. International Union of Crystallography. Kluwer, Dordrecht.

Samuels LE (1992) *Optical Microscopy of Carbon Steels.* American Society of Metallurgists.

Young RA (1993) *The Rietveld Method.* International Union of Crystallography. Oxford University Press, Oxford.

Chapter 16
The guns: firearms of the Kelly gang and police

Malcolm Dodd and Craig Cormick

Despite having stolen many more sophisticated firearms, the Kelly gang actually went into the siege at Glenrowan with fairly antiquated guns, and were severely outgunned by the police.

The firepower of the Victorian police of the Kelly era was sophisticated for its time, the force having repeating rifles, shotguns and pistols at its disposal.

The shotguns of the day were essentially very similar to those we see today (absent the arrival of pump-action models). Single- and double-barrel shotguns were readily available, firing both lead and iron-type shot pellets of varying size, depending on the intended use. The discharge could be lethal at close range and could certainly be disabling over distance.

The pistols available to the police comprised Webley and single-action Colt revolvers. The Webley had a very sensible break-open action, allowing the pistol to chamber six 0.455 inch calibre cartridges, and it fired its bullets at a velocity of 183 m per second. The Colt Army pattern revolver could also chamber six 0.45 inch caliber cartridges, and fired at a similar velocity. Both firearms were accurate up to 25 m.

The Spencer rifle was, however, the 'jewel in the crown'. This advanced firearm, patented in early 1860, literally changed the outcome of the American Civil War, with its great accuracy over distance and its seven-cartridge magazine able to be discharged rapidly. This lever-action rifle measured 1194 mm in length and was able to discharge seven 0.56 inch (14.2 mm) projectiles with ease. A skilled shooter was capable of discharging 12 rounds per minute.

The muzzle velocity was 366 m per second and it was therefore regarded as a high-velocity weapon. Bullets fired from it were capable of inflicting lethal injury over great distances.

The Snider rifle, although able to chamber only a single round, was also regarded as highly accurate. A breech-loading firearm, discharging a 0.557 inch (14.6 mm) projectile that travelled at 355 m per second, it did suffer from the occasional misfire. The cartridges of the day consisted of a paper cylinder with a brass base and primer. These could buckle in the breech, requiring manual removal before the next round could be chambered.

The Kelly gang's guns

The Kelly gang, by comparison, had a mix of weapons, some quite antiquated and some more modern, stolen from the police. Ned Kelly's own preferred weapon had long been an ancient 0.577 inch calibre carbine that had both its butt and barrel sawn off so it was only about 60 cm in length. Also, notably, it was bound together with waxed string.

This is the gun that Ned Kelly used to shoot the police at Stringybark Creek.

As the four policemen involved in that attack had come very heavily armed, the Kelly gang acquired quite an arsenal of weapons. They now had four Webley revolvers, Constable Scanlon's 0.500 inch calibre seven-shot Spencer carbine rifle and Sergeant Kennedy's double-barrelled shotgun. Both of these had been borrowed by the police, Scanlon's from a gold escort and Kennedy's from Reverend Sanderford of Mansfield. The latter gun, the shotgun, was said to have been the one that was used to later kill Aaron Sherritt in the lead-up to the siege at Glenrowan.

After robbing the bank at Euroa in December 1878, the gang added two rifles, two double-barrelled shotguns, eight revolvers and a Snider-Enfield 0.577 inch calibre rifle to their collection. This last gun, taken from a hunting party that the gang bailed up, was nicknamed 'Betty' and became Ned's favourite gun for a time. He even carved the letter 'K' (for Kelly) onto its butt.

The gang's next robbery was at Jerilderie in New South Wales and they added to their collection four revolvers and two 0.539 inch caliber carbine rifles.

They also later acquired a police Martini-Henry rifle that was said to have been used to test the strength of their armour against police bullets.

It is estimated that the Kellys had around 30 firearms by the time they were ready for the showdown at Glenrowan. However, most of the best guns were believed to have been given to sympathisers to use and Ned returned to his old string-bound sawn-off carbine. He also had a revolving Colt rifle and two revolvers – one of which was a Navy Colt revolver that had been taken at Jerilderie (see Figs 16.1, 16.2). This gun needed to be loaded with fiddly percussion caps and balls, and was the one that he banged against his armour when confronting the police before he was shot down.

Because Ned Kelly carried an assortment of guns he needed an assortment of bullets, balls and percussion caps, which he carried in a leather pouch. However, after being wounded in the arm and then the hand in the gun battle, Ned had great trouble working his guns and was certainly unable to handle the complicated percussion caps and balls: he was overheard by Constable Phillips to say to Joy Byrne inside the Inn, 'What are you doing there; come with me and load my rifle. I am cooked.' To which Joe replied, 'So am I. I think my leg is broke.' To which Ned replied, 'Leg be damned; you got the use of your arms. Come on; load for me. I'll pink the buggars'(*Argus*, 1881).

Joe Byrne was later hit in the groin, the police bullet finding a gap in his armour, and bled to death. Ned was felled by 'duck shot' fired at his legs by Sergeant Steele, when he advanced on police lines the following morning. Sergeant Steele later souvenired Ned's leather ammunition pouch.

After his capture, Ned was reported to have said to his sisters who came to see him:

> I was at last surrounded by the police, and only had a revolver, with which I fired four shots; but it was no good. I had half a mind to shoot myself. I loaded my rifle, but could not hold it after I was wounded. I had plenty of ammunition, but it was no good to me. I got shot in the arm (*Argus* 1880).

Several of the Kelly gang's guns have survived; a pistol that belonged to Dan Kelly, with his name and the year 1876 engraved into the handle, sold at auction in 2012 for $122 000. Also, Ned's revolving Colt rifle, which is held by the same private owner who holds Joe Byrne's armour, Rupert Hammond, was tested against one of the percussion caps found at Glenrowan and was determined to be a match, confirming the gun had both been at Glenrowan and had been fired by the outlaws there.

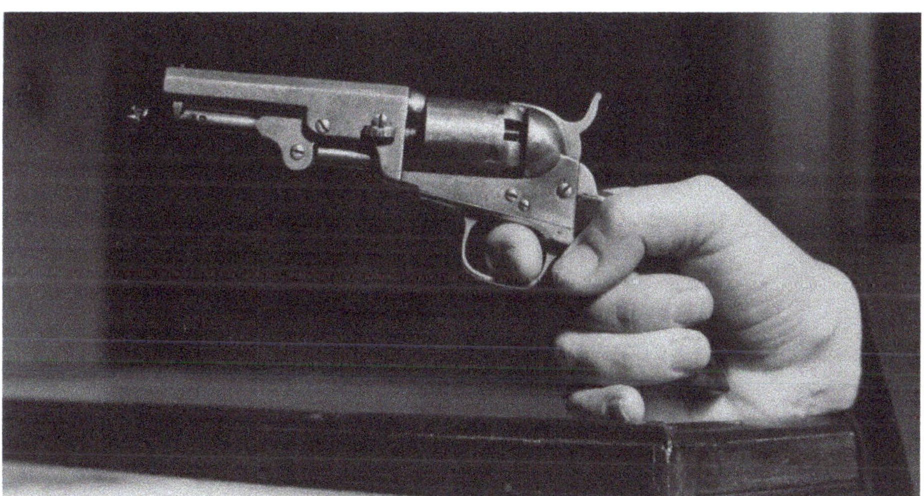

Fig. 16.1: Ned Kelly's Colt pistol. State Library of Victoria.

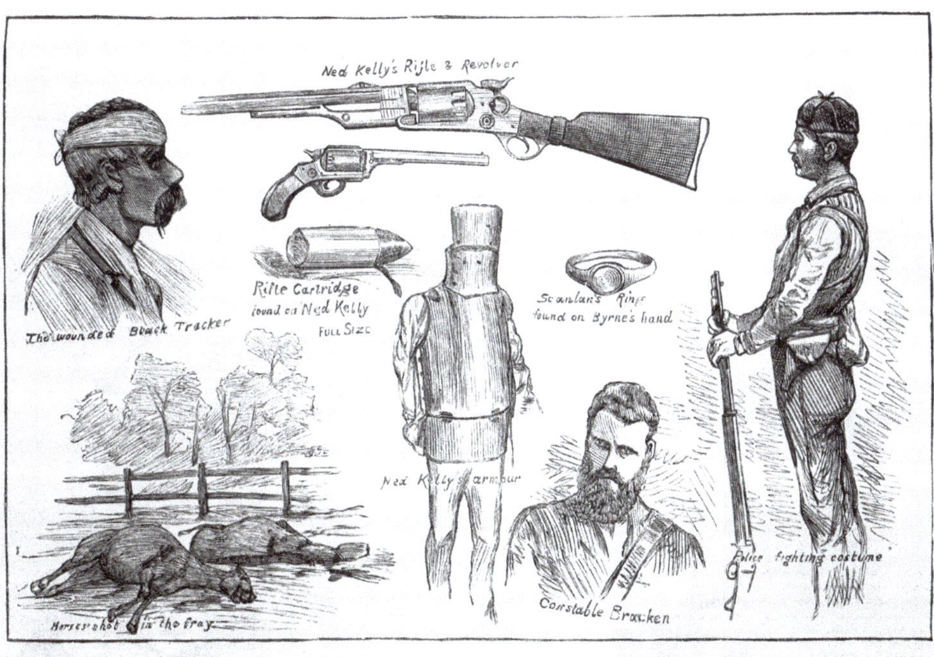

Fig. 16.2: Incidents sketched at Glenrowan, from the *Illustrated Australian News*, 3 July 1880, showing Ned Kelly's rifle, revolver and rifle cartridge, as well as the guns used by police. State Library of Victoria.

References

Argus, Tuesday 29 June 1880.

Argus, Tuesday 21 September 1881. http://www.ironoutlaw.com/html/weapons.html (accessed on 4/12.2013).

Chapter 17
Ned's injuries and their treatment: then and now

Frank McDermott and Max Esser

Ned Kelly sustained several gunshot wounds at Glenrowan, causing significant injury to his leg and arm. This analysis of his treatment looks at whether it was consistent with best practice at the time, and also how he might have been treated today.

Ned Kelly sustained several gunshot wounds to his limbs in the firefight with police surrounding the Jones Public House at Glenrowan on 27 June 1880. The police were well armed with Spencer and Snider rifles, Webley and Colt revolvers and shotguns.

Ned Kelly escaped the inn at dark and returned to attack the police from their rear the following morning. He was initially seen wearing a long grey overcoat and iron mask and firing with a revolver while walking from tree to tree. Although the police fired at him persistently he seemed bullet-proof. Guard Dowsett exclaimed, 'By God, it is the devil.' After taking another shot at him Senior Constable Kelly replied, 'No, it must be the bunyip' (*Argus*, 2 July 1880).

Sergeant A. Steele, however, believed that Kelly was encased in metal mail armour and aimed at his legs. The shots felled Kelly and, with the assistance of other police, he was disarmed and secured. The place where he fell was saturated with blood and a considerable amount of blood was found on the ground by an adjacent tree.

A 0.45 Martini Henry bullet had penetrated Ned 's bent left elbow, inflicting wounds below and above the elbow (Jones 2003), and he was also struck by a bullet ripping into his right foot near the great toe and passing through to the sole near the heel. As a result he was almost totally disabled.

In a letter written by Kelly from Melbourne Gaol to the Victorian Governor, the Marquis of Normanby, he stated that he received those injuries the prior evening (PROV,VPRS 4966, 1880).

Inspection of Ned Kelly's armour, which had been forged from a stolen ploughshare, found the armour to be 6 mm thick and to weigh 44 kg. There were apparently five bullet marks in the helmet, three in the breast plate, one on the shoulder plate and nine on the back plate (*Argus*, 29 June 1880). Although it had not protected him from being shot in the arms and legs, the armour had protected him from life-threatening head, chest and abdominal gunshot wounds.

On 7 August 1880 at the Beechworth Court committal hearing, about six weeks later, Ned Kelly was reported to have limped into the dock and was described as remaining lame and maimed (*Argus*, 7 August 1880).

Description of injuries

Left arm injuries

Dr A. Shields, chief medical officer and surgeon of Melbourne Gaol, writing the official medical report, described two wounds caused by a bullet going through the fleshy part of the left lower arm, one above and one below the elbow. The two openings were believed to be caused by a bullet traversing the arm when the elbow was bent. Ned Kelly agreed that his elbow was flexed when it was struck, and thereafter he could not hold a rifle (*Argus* 29 June 1880).

Dr C. Ryan, who attended Ned Kelly on the train to Melbourne, also found that his left arm was pierced by two bullet holes, one above and one below the elbow (*Argus*, 30 June 1880).

A large bullet wound 9–10 cm above the elbow on the lower surface of the arm was also described by Dr J. Nicholson of Benalla. He stated that the bullet had not exited and was lying somewhere in front of the elbow joint (*Argus*, 2 July 1880).

Forensic examination of Ned Kelly's remains was undertaken in 2009 by Dr S. Blau at the Victorian Institute of Forensic Medicine (Blau 2009), discussed in detail in Chapter 5. It found, among other things, that the proximal or upper parts of the bones of the forearm had been removed post mortem. The examination also found a lesion 29 × 14 mm on the lower inner-end, or 'medial aspect', of the left upper arm bone (humerus) that was associated with new growth of bone after an injury.

The report of the CT scan of Kelly's remains in June 2009 discusses this injury and mentions that the the injury of the lower arm was adjacent to and could have injured the ulnar nerve. This would have led to subsequent weakness and wasting of the left arm, and withering of the left hand (*Herald*, 10 October 1880). His left hand was reported as having a wasted and crushed appearance (Jones 2003).

A photograph of Ned Kelly taken at Melbourne Gaol at his request, by Charles Nettleton, shows that his withered left arm has been masked by him holding the chains attached to his leg irons (Jones 2003) (see Fig. 17.1).

This same injury was described in a report prepared in 2011 by Dr M. Dodd, Senior Forensic Pathologist at the Victorian Institute of Forensic Medicine. The description was in accordance with that of Dr Blau, and he concluded it was associated with a creasing injury from a bullet with subsequent new bone growth (see Fig. 17.2).

Right foot gunshot injury

The official medical report of Dr A. Shields states that the projectile track was marked by two openings, one on the top of the ball of the great toe, and the other on the sole of the foot. A bullet had passed in a slanting direction backwards, terminating in a long slit-like wound near the heel (*Argus*, 30 June 1880). There was unspecified bone injury.

The forensic anthropology report of 2009 of Ned Kelly's remains showed new bone formation associated with a fracture extending across the main joint of the big toe (Blau 2009).

The CT scan of 2009 revealed metallic debris and bony fragments present, that were characteristic of a gunshot wound.

Fig. 17.1: Portrait of Ned Kelly taken in Melbourne Gaol by Charles Nettleton, showing him hiding his withered left hand and arm by holding the cord attached to the leg iron. Ian Jones.

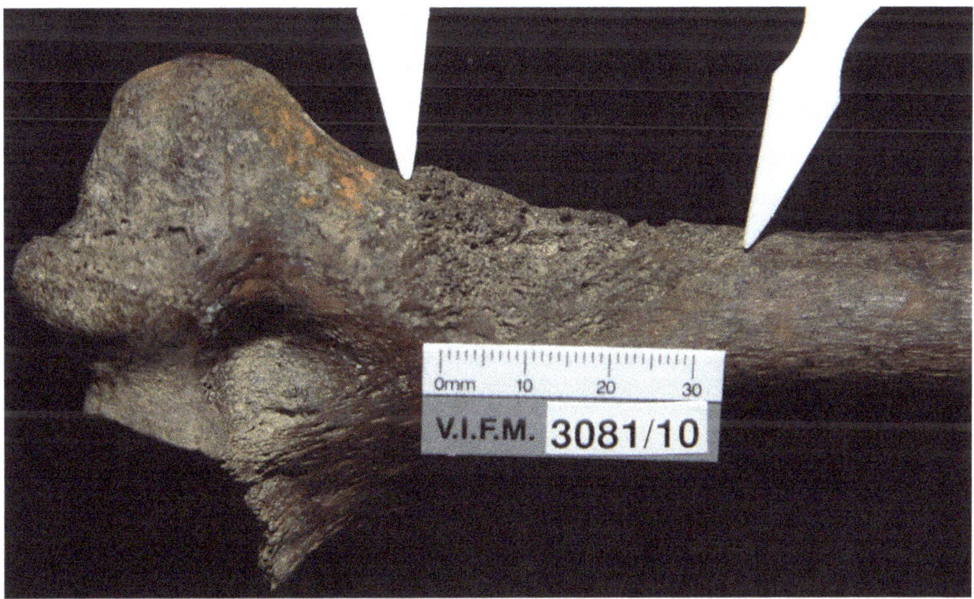

Fig. 17.2: Photograph of Kelly's humerus bone of the upper arm, with callus indicated, close to the location of the left ulnar nerve. Victorian Institue of Forensic Medicine.

Right hand injury

Two wounds were present in the right hand. One shotgun slug had entered the ball of the thumb and a CT scan of the right hand showed that the carpal bones of the wrist were missing and that only three of the five large metacarpal bones of the hand were present (the missing bones were most likely left behind from the exhumation at the Old Melbourne Gaol).

Right thigh, lower leg and groin injuries

There were several shotgun slug wounds on the outer side of the right thigh and lower leg. Dr C. Ryan had removed one of eight or nine shotgun pellets and stated the wounds were not considered to be dangerous. It was reported, however, that Ned Kelly complained of slugs (or pellets) in his lower leg and had pain about the knee (*Age*, 16 August 1880). The forensic anthropology review of his remains conducted in 2009 showed a circular hole, or perforation, 8.08 × 7.6 mm in external diameter on the outer side of the right lower leg bone (tibia) (see Fig. 17.3). The 3D CT reconstruction scan showed two projectile remnants in the bone around the single entry wound.

One of these metallic balls was actually moving freely inside the tibia and the other appeared fixed. Associate Professor David Ranson and Dr Richard Bassed were successful in extracting these projectiles at the Victorian Institute of Forensic Medicine, and found they comprised spherical metal discs that were dark brown in colour and measured 7.37 × 5.62 mm and 5.61 × 5.68 mm, respectively. The objects were consistent with flattened shotgun ball projectiles, and it appeared they had quite coincidentally entered the same hole (see Fig. 17.3).

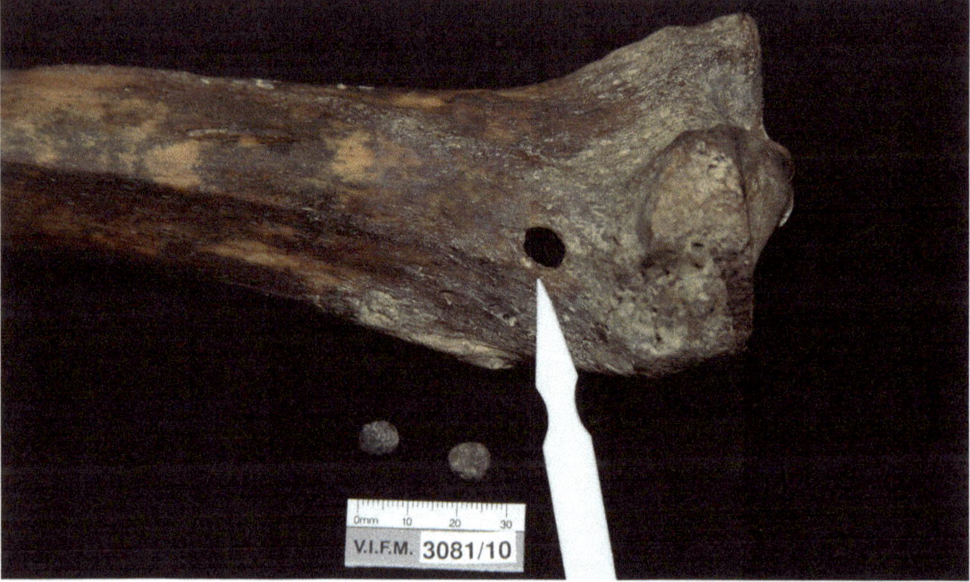

Fig. 17.3: Ned Kelly's upper right tibia bone of the lower leg, showing a circular perforation. Also shown are the projectiles retrieved from the right tibia. Victorian Institue of Forensic Medicine.

Medical management

The available reported information on Ned Kelly's medical management is scant, with records now lost or unavailable. After capture it was reported that Kelly appeared to be suffering from severe exhaustion and was described as cold, white and smelling of brandy. He also complained of coldness of the feet. Sometime after his capture he was carried to the Glenrowan Railway Station and laid on the floor of the guard's van (see Fig. 17.4) before being moved to the Station Master's office. His wounds were dressed by Dr Nicholson, of Benalla.

Kelly was then taken by train to Melbourne, under the care of Dr C. Ryan. His pulse en route was noted to be raised to 145 beats per minute. Although described as feverish, his temperature is not available. When taken from the train he was moved by wagonette to Melbourne Gaol. His pulse there was measured at 114 beats per minute. At North Melbourne Railway Station spectators noted that his hands and feet were bound up(see Fig. 17.5).

The *Age* of 16 August 1880 reported that Dr Shields extracted, with difficulty, a bullet firmly embedded in the muscles of the foot, although it was probably a fragment of bullet.

Fig. 17.4: Sketch of Ned Kelly, showing wounds to his foot and arm, and bruises to his face from his helmet, while laid in the Station Master's house. Drawn by T. Carrington for the *Australasian Sketcher*. State Libary of Victoria.

Fig. 17.5: Ned Kelly's arrival in Melbourne, 3 July 1880. State Library of Victoria.

Dr Shields, the Medical Officer at Melbourne Gaol, considered that Kelly's condition was favourable enough and that a speedy recovery was expected. On 26 August he was removed from the gaol hospital to a cell.

Wound treatment at the time

Was Ned Kelly's wound treatment consistent with the approaches favoured at that time? He sustained multiple, non-life-threatening, low-velocity gunshot wounds to his extremities, from both rifle and shotgun fire at Glenrowan. The available accounts of his treatment are limited to that received in Glenrowan, en route to and initially at Melbourne Gaol.

Unfortunately the Melbourne Gaol hospital records related to his treatment are not available, nor are there any records of the post mortem dissection that occurred in 1880. Information is also lacking in regard to the left arm injury. Nonetheless, it is reported that this resulted in limited movement of the elbow and marked muscle wasting. It is unknown whether the bullet fractured neither, one or both of the forearm bones and/or injured the elbow joint. However, the lesion found on the humerus bone is consistent with a bullet creasing it, alongside where the ulnar nerve runs. It is probable that this injury caused ulnar nerve damage, with resultant wasting of the limb. The nearby median nerve may also have been injured.

Although bone damage was reported in the right foot injury, no specific diagnosis was made before examination at the Victorian Institute of Forensic Medicine which detected a

Early treatment of gunshot wounds

The earliest treatment of gunshot wounds included wholesale blood-letting, wide opening or dilatation of the wound with extensive probing to remove the projectile and foreign material, encouraging bleeding to stop by using boiling oil, and liberal use of amputation for complex wounds and fractures (Swan and Swan 1989).

Using his experience at Bell Isle during the Seven Years War, John Hunter differed dramatically from the conventional radical treatment of gunshot wounds at that time by advocating conservatism, allowing the wound to suppurate and then slough off the dead material (Hunter 1794). He advised against aggressive surgery for smaller wounds, his observations and recommendations based on low-velocity pistol and musket wounds. He also recommended exploring the entry point of all gunshot wounds, but advised that the wound should only be opened if the artery was injured, if bones were fractured or if foreign objects were present. He relied on natural suppuration to effect healing and believed that doctors cutting away dead and damaged tissue only increased inflammation.

During the US Civil War (1861–65) limb amputation was commonly performed for complex limb wounds and open fractures, supported by the use of tourniquets and cauterisation to control bleeding – gradually this was replaced by ligatures (sewing or clamping blood vessels closed). Conservative treatment was employed in cases of gunshot wounds involving soft tissues or minor bone fractures and bullets were removed only if within easy reach of the surgeon (Manring *et al.* 2009.) The wound was cleaned of bone fragments, clothing and other debris and dressed with bandages.

This type of conservation left the limb intact rather than amputating it, but the use of unsterilised instruments, unwashed hands and dirty bandages often introduced infection (NMHM 2013). Any open wound almost always became infected and these infections often resulted in gangrene and death.

Lewis S. Pilcher, in 1883, only three years after Ned Kelly's death, advised that for wounds of considerable extent, the skin should be cleaned thoroughly with soap and water and a brush and razor, and carbolic acid applied (Pilcher 1883). He also advised that the existing wound opening could be enlarged with a knife to allow removal of coagulated blood and foreign material.

Frank McDermott and Max Esser

fracture of the bones of the foot (proximal first phalanx involving the first metatarsophalangael joint) (see Chapter 5). It would have been inflicted by the rifle gunshot wound to the ball of the great toe.

Ned Kelly's treatment was, of course, delivered before the introduction of modern medical techniques such as radiographic imaging, tetanus immunisation and antitoxin, intravenous fluid resuscitation, blood transfusion, antibiotics, identification of bacteria and recognition of their pathological consequences, the use of full sterile surgical technique, improved methods of fracture fixation and stabilisation and reconstructive surgery.

Also, several important surgical principles were not generally established until World War I, over 30 years later. These included removing dead or infected tissue, widening of the wound, incisions in the tissue covering muscle to prevent muscle death from swelling or strangulation, and delayed wound closure to reduce the risk of wound infection.

Gunshot wound management before and around 1880

A literature search was made for surgical textbooks and articles related to gunshot wounds and care before and around 1880, to determine how Ned Kelly's treatment met the standard practices of the day. A popular surgical textbook, appearing in 1860 (Erichsen 1860) and in frequent new editions including the sixth in 1877, was written by J.E. Erichsen, Surgeon and Professor at University College Hospital, London. In it, gunshot wounds were described as a variety of lacerated and contused wounds 'often with entry and exit projectile orifices'.

The bruising, or contusion, was related to the velocity and force of the projectile, and small shot fragments that lodged in the tissue under the skin may have led to infection with pus formation (suppuration). Gunshot wounds infrequently involved serious haemorrhage or bleeding from major vessels, but caused inflammation leading to suppuration sometimes with deep tissue strangulation. It also stated that dead tissue (slough) begins to separate from 12 to 20 days after the injury.

Wounds of that era were at high risk of complication by bacterial infections, with subsequent hospital gangrene or late bleeding. Amputation was advised for gunshot wounds where there was considerable tissue damage.

The surgical principles of the time involved extraction of foreign bodies and unattached bone fragments using a finger or forceps. Incisions were made over small shotgun pellets, which were then extracted to lessen the risk of infection. Additional cuts or incisions through connective tissue covering muscles were commonly required to prevent the muscles from dying and were followed with wound irrigation, or washing.

This minimal approach seemed satisfactory for non-civilian injuries and for gunshot wounds of low-velocity even when bone was fractured. Ned Kelly's wound treatment was in keeping with this minimal conservative approach.

Current treatment of Edward Kelly's injuries

If Ned Kelly had been treated for his injuries today, the approach would be quite different.

Careful assessment of the patient, Edward Kelly, at the accident scene would include assessment of his airway, breathing and circulation. If necessary he would be resuscitated with intravenous fluids and painkillers, such as intravenous morphine.

Moved to the hospital emergency department, his wounds would then be inspected for size, shape and degree of contamination, and both the entry and possible exit wounds of bullet injuries would be noted. Photographs of wounds would be taken by local medical staff for surgeons off-site to facilitate planning and surgical removal of damaged tissue.

A temporary dressing would then be applied, to minimise the risk of cross-infection, and antibiotic would be administered intravenously with tetanus toxoid and anti-tetanus antibodies, considering the bullet wounds had been sustained in a tetanus-prone, agricultural environment. His limbs would then be splinted and X-rays taken from several angles.

Fractures adjacent to joints may have required CT scans with 3D reconstructions to accurately define their extent.

On the day of admission, cleansing the wound would be performed in the operating room under a general anaesthetic, with a tourniquet applied. All contamination, such as dead skin, fragments of clothing, dirt, bullets or other foreign bodies would be removed. Healthy, clean non-damaged skin edges would be retained and a machine used to wash out the wounds with a sterile solution.

Any muscle injuries associated with the passage of a bullet would require release of the overlying connective tissue (fasciotomy). X-rays and CT scans would be examined to see if any affected joint needed irrigation or washing.

In addition, blood vessels and nerves affected by the track of a projectile may have had to be explored, with the vessels being repaired and the nerves tagged for later repair. Fractures of bone would be treated with internal or external fixation (McRae and Esser 2008) with reconstruction of more complex fractures being planned for subsequent surgery, one to five days later.

A negative pressure dressing (vacuum-assisted) (Blum *et al.* 2012) would be applied to the wounds and they would be re-examined after 24 hours, with repeat removal of dead and damaged tissue considered. A planned delayed closure of the wound itself would occur between day two and day five, following the injury.

Specific management of injuries sustained by Edward Kelly

Left elbow injury

Initial X-rays would have defined the presence of any fracture and involvement of the elbow joint, as well as the presence of bullets. The wound would have been cleaned and the adjacent ulnar nerve explored. The elbow joint would very likely have been damaged, and hence washed out at the time of initial surgery. Foreign bodies such as bullets would have been removed. Other parts of the arm may have been involved in this injury and would have also required exploration, with a view to repair at a later time.

Any other bone damage would have been identified on a CT scan, and if significant bony instability was present an external fixateur, using metal pins and a frame, would have been used to provide temporary stability. Complete stabilisation of any fracture involving joint surfaces would be performed either straight away or at the time of delayed closure of the wound, two to five days later. A negative pressure dressing would also be applied and the wound would be reassessed for additional removal of damaged tissue and washing after 24 hours.

Right hand injury

X-rays and CT scans taken before surgery would have defined the degree and extent of injury. The wound would then be opened, dead and damaged tissue removed and a negative pressure dressing would be applied and the hand placed in a splint. Stabilisation of the damaged bones would need to be considered if a fracture or dislocation of the bones of the hand had occurred.

Right thigh and groin injury

These wounds would have been explored and dead and damaged tissue removed, then they would be washed and left open. The artery and vein the in the leg may have required investigation for damage and, if the wound was extensive, a negative pressure bandage would be applied. The wound would be reviewed after 24 hours for possible repeat removal of damaged tissue and washing, with a view to closing the wound between two and five days after the injury.

Right lower leg injury

The entry point of the bullet would have been very close to a significant nerve of the leg and this would have to be inspected when the wound was examined and damaged tissue removed. Depending on the trauma found, tissue covering muscle may have been cut to relieve the build-up of pressure (fasciotomy). A CT scan, with 3D reconstruction, would have defined the location of bullet fragments. These could have been removed at the time of initial surgery with the assistance of real-time X-rays. The entry wound, close to the back of the knee joint, may have caused damage to the blood vessels there and further investigation of them would be considered.

The stability of the knee would have been assessed and repair of damaged ligaments would have been undertaken then, or at the time of closure of the wounds. A topical negative pressure dressing would have been applied and the knee splinted.

Right foot injury

The wound would have been treated like the others, then a local or a closed vacuum dressing would have been applied. A supportive structure for the foot and ankle, known as a 'backslab', would have been used and major bony fragments would have been reduced and then stabilised with special wires and/or pins. Definitive fixation may have had to be performed by the time of closure of the wounds, again two to five days later.

Historical perspective

Ned Kelly's use of metal armour was successful in protecting him from life-threatening head, chest and abdominal injuries, as all his injuries were limb injuries. These maimed and disabled him, but were not life-threatening. His treatment at the time, using a conservative approach, was consistent with methods of the time and did not represent any neglect of his care.

The aim of modern wound management, by comparison, is to achieve a clean wound with viable, uninjured wound edges and healthy adjacent tissues that will facilitate healing. Early movement and return to function of the injured limb is also more readily achieved with modern techniques of bone stabilisation, such as the use of plates and screws (internal fixation) and/or external fixation. Appropriate physiotherapy and rehabilitation are further essential components of modern injury management.

The consequences of Kelly's injuries would have been dramatically improved by contemporary surgical management as compared with those provided by the conservative minimal approach favoured in 1880. Under modern medical intervention he could have expected healing of his wounds and better function, even in the short period of five months, before his execution on 11 November 1880.

Acknowledgements

The authors wish to thank Victoria Winship, Manager of Information and Data Analysis, Victorian Institute of Forensic Medicine, and Kaye Lionello, Cabrini Medical Centre, for secretarial assistance. We also thank Dr M. Dodd, Victoria Winship, Fiona Leahy and Mellita Hunter, all of the Victorian Institute of Forensic Medicine, for critical review.

References

Age, 16 August 1880.
Argus, 16 August 1880.
Argus, 29 June 1880.
Argus, 30 June 1880.
Argus, 1 July 1880.
Argus, 7 August 1880.
Blau S (2009) Forensic anthropology and police investigations. *Australian Police Journal* **63**(1), 8–14.
Blum M, Esser M, Richardson M, Paul E, Rosenfeldt F (2012) Negative pressure wound therapy reduces deep infection rate in open tibial fractures. *Journal of Orthopaedic Trauma* **26**, 499–505.
Erichsen JE (1860) *The Science and Art of Surgery*. Vols 1 and 2. James Walton, London.
Herald, 10 October 1880.
Hunter J (1794) *A Treatise of the Blood, Inflammation and Gun-shot Wounds*. Nicol, London.
Jones I (2003) *Ned Kelly: A Short Life*. Revised edn. Hachette Australia, Melbourne.
Manring MM, Hawk A, Calhoun JH, Andersen RC (2009) Treatment of war wounds: a historical review. *Clinical Orthopaedics Related Research* **467**, 2168–2191.
McRae R, Esser M (2008) *Practical Fracture Treatment*. 5th edn. Churchill Livingstone, Edinburgh.
National Museum of Health and Medicine (NMHM) To bind up the nation's wound. http://warmedicalmuseum.
Pilcher LS (1883) *Treatment of Wounds, its Principles and Practice General and Special*. William Wood and Co., New York.
PROV (1880) VPRS 4966/PO, Item 4, Record 12: letter by Ned Kelly while incarcerated to his Excellence the Marquis of Normanby re: Glenrowan.
Swan KG, Swan RC (1989) *Gunshot Wounds: Pathophysiology and Management*. Yearbook Medical Publishers, Chicago.

Chapter 18
Sifting through the past: the archaeological dig at Glenrowan

Adam Ford

An archaeological examination of the inn where Ned Kelly and his gang had their last stand at Glenrowan reveals some unique insights into how the gang fought during the battle.

The Kelly story, particularly the siege at Glenrowan – or Ned Kelly's last stand – is unique in Australian history in that it is well known not just across the whole country, but across the world. But the story has evolved from historical fact to myth and legend, and has a polarised following. To some, the gang members, particularly Ned Kelly, are considered to be chancers, thieves and cold-blooded murderers whose rhetoric about class and Catholic Irish persecution were only an attempt to justify their crimes. Yet to many others he is a hero of the down-trodden and the champion of the poor.

The iconic image of Ned returning, terribly wounded, into the fray in an attempt to rescue his brother and friend is the quintessential picture of the cherished Australian ideal of mateship, and was a story greatly exploited in World War I.

Without the siege the Kelly gang may have slipped into history as a bunch of bloodthirsty young bushrangers and sat alongside the likes of Harry Power. But the siege happened, and at the Glenrowan Inn the Kelly legend was born.

In 2008 the Rural City of Wangaratta commissioned DIG International Pty Ltd to conduct the first archaeological excavations at the historic site of the Glenrowan Inn. Through a regional development grant the city, in collaboration with Heritage Victoria, proposed the excavation as part of a profile-building and rehabilitation strategy for Glenrowan and the region.

The aim of the project was therefore two-fold: to conduct an archaeological investigation of the site to discover the physical remains of the siege (if possible) and to carry out a comprehensive public and media engagement program.

The inn site is part of an area described as the Glenrowan Heritage Precinct, which lies at the centre of the Glenrowan township and is considered one of the most historically significant landscapes in Australia. This is reflected in the protection of the precinct by both federal and state legislation, its recording on the National Heritage List under section 3 of the *Environment Protection and Biodiversity Conservation Act 1999*. The list includes sites of outstanding national heritage to the Australian nation. The area is also recorded as a site of state heritage significance on the Victorian Heritage Register (register number H2000) under the *Heritage Act 1995* (Vic.).

Developing a research design for the dig

Archaeological excavation is an unrepeatable experiment – it destroys sites and deposits as they are dug – so it is imperative that excavation methods are defined by a considered research design. However, to develop a research design for the Glenrowan project the archaeological challenges of the site had to be better understood.

For instance, the land-use history of the site included activities before and after the siege, so one of the challenges of the project was to identify contexts and artefacts that specifically related to the events of the siege in June 1880, an intense but brief 12 hours or so.

However, because ceramics and glassware of the period would have decoration and other attributes also common to times before and after the siege it was not expected that finding such artefacts would be useful in defining different occupation phases – or times of occurrence. We therefore decided that detailed observations of artefact location, nature, condition and association with soils and features would be of greater use in determining their chronological and activity sequences.

The following 14 research questions were proposed as terms of reference to guide the project's approach and resources.

1 Were there physical remains of the siege, and if so could we make any sense of them?
2 What could these remains tell us about the events that could not be found elsewhere?
3 Could the archaeological data be used to verify or contend with official accounts of the siege?
4 Could the archaeological data give us more information on the life of Ann Jones, proprietor of the inn?
5 Could the archaeological evidence be used to interpret the site in the future?
6 Could the archaeological excavation show the extent and level of disturbance caused by souveniring at the inn site immediately after the siege?
7 Could the archaeological excavation show the extent and level of disturbance caused by souveniring at the inn site in the last 130 years?
8 Could the archaeological excavation observe the elements that created the site, both natural and human?
9 Could soil sieving and soil chemical analysis identify where Dan Kelly and Steve Hart were partially cremated during the fire that ended the siege?
10 Could we prove or disprove the theory that Dan Kelly and Steve Hart escaped from the siege?
11 Could analysis of the artefacts give an insight into the life of the inn's proprietor and other locals of Glenrowan?
12 Could we, through careful mapping, recreate the direction and progression of the battle?
13 Using forensic ballistic analysis, could we identify individual weapons and, if so, their direction and movements?
14 Would we be able to plot the positions of the Kelly gang over time from artefact remains such as cartridges, or groupings of return fire ballistics?

Approach

Our major focus needed to be on identifying structural remains, together with the associated soil deposits, so establishing the outline of the original inn, and if possible the individual rooms, was a primary goal. This was seen as important if the project was to overlay the known or recorded historical events and time sequence to the artefacts found. Archaeological practice requires that all the data found must be recorded as objectively as possible although it is often interpreted subjectively later. However, this has long been acknowledged as problematic as it does not take into account the subjective and complex interrelationship of the archaeologist to the archaeology at the time of excavation. It encourages a dry and two-dimensional interpretation of the human condition, which is anything but dry and two-dimensional (particularly in this scenario).

And clearly this approach would not work at the Glenrowan Inn site, as it had too much history and notoriety. Therefore it was proposed to adopt a more reflexive approach, still intending to recover and record most data in a traditional manner, but acknowledging that discussion and consideration of the findings would encourage more creative analysis of them. Indeed, understanding what the team were thinking about as they recorded the site benefited our interpretation of it.

The team also had to be mindful of 'fuzzy stratigraphy' when excavating and recording the site, where objects have been moved out of their original context, as would have happened during the mass souveniring immediately after the fire at the end of the siege.

The existence (or not) of 'ghost' contexts was also considered. Ghost contexts are elements of the site that must have been there originally but have been removed over the years or immediately after the event (e.g. munitions). Many sites have been raided in the past, and in this instance it was considered appropriate to incorporate these missing elements into the primary record, as the munitions were a significant component of the siege.

Third, 'invisible archaeology' would also need to be included for interpretation of the site during and after the dig – that is, actions that occurred but no actual trace of them could be found.

The siege was a small but significant conflict and was arguably the last major battle on Australian soil. As such the site was regarded by the archaeologists as a battlefield.

While this did not require any specific variations to methods of excavation, the following was kept in mind:

- people died at the site as a result of the siege and there were living relatives in the local community, and therefore the project had to be approached with sensitivity and in a balanced scientific manner;
- battlefield deposits (if they exist at all) are created very rapidly and, as mentioned above, create fuzzy contexts;
- rational, planned activities don't occur during battles, so rational reasoning for interpretation of remains may not be appropriate.

In summary, the site had five dimensions which all had to be recorded or at least borne in mind. The normal three dimensions could be covered by the standard recording strategy, the fourth dimension, time (in the past both linear and non-linear) was to be incorporated into the day-to-day interpretation of discoveries through a thorough understanding of the siege timeline and discussion. The fifth dimension in this case was the excavator and his/her bundled subjectivity about the siege story and Ned Kelly.

Historical research

I had undertaken extensive historical research for earlier archaeological investigations at the siege site, however, for this project professional historian Alex McDermott carried out the principal historical research. He had three aims:

1 to identify and locate all primary and secondary historical documentation that described the siege, the inn and movements of the Kelly gang and police;
2 to construct a clear picture of the land use history of the site, including periods before and after the siege;
3 to establish the archaeological potential of the site based on the land use history and the possible nature of the archaeological deposits based on historical descriptions.

Of course, considerable historical research of the Kelly story has been carried out over the years. However, there has not been a study to specifically look at the archaeological potential of the inn site from evidence in the historical record. So, while there is a significant collection of historical accounts of the events and lives involved, there is scant secondary information about the construction of the inn, or what the inn contents may have been like.

Excavation method

Archaeologists excavated the ground by hand and recorded findings by assigning context numbers to deposits, soil and structural features as determined by the site director and site supervisor. Each finding was described on a special sheet and recorded photographically and on scaled plan drawings.

All artefacts found were recorded to where and how they had been found and were bagged and labelled, then catalogued and retained, in accordance with Heritage Victoria's guidelines. Special finds (as defined by the supervising archaeologist) were photographed where they had been found, and the locations were recorded in detail.

Results: historical research

Glenrowan was originally founded roughly 1.5 km west of its present location but the focus of the town shifted with the relocation of the railway station in 1875. This was due to the fact that north-bound trains had difficulty starting on an incline at the original siding. Moving the station allowed trains to stop on the level, at the crown of the pass through the Warby Ranges.

In 1873 Ann Jones bought a cleared block on which a timber slab hut and stables already stood. Ann lived in the hut and it is thought she operated a lodging there before the construction of the Glenrowan Inn.

In 1878 Ann Jones commissioned local builders, Emery and Jarvis from Wangaratta, to construct an inn to the south of her residence. The original plan was for a three-room, single-gabled building, but she requested two bedrooms to be added in a lean-to, skillion-

Who was Ann Jones?

Ann Jones was born Ann Kennedy, in Tipperary, Ireland in 1833 and came to Victoria in about 1854. She is recorded as having married a Welsh labourer, Owen Jones, in September of that year. The couple moved around many places in Victoria and she bore 11 children (although several died while young) before setting up her inn at Glenrowan.

She had previously run a tea room (more likely a sly-grog shop) in Wangaratta but in 1879 took a loan, had the Glenrowan Inn built and obtained a licence to sell liquor. She picked a good location, just opposite the newly moved railway station, and there was only one other inn or hotel in the township to compete with.

Ann Jones fitted her inn out well, perhaps looking for a more upmarket clientele than those at McDonnell's Hotel on the other side of town, which Ned Kelly and his friends occasionally visited.

There is reason to believe that Ned wanted to keep a close eye on her. Not only had she been reportedly a policy spy, but a few years earlier she'd taken Ned's mother Ellen to court over an outstanding debt of £2 (Wright and McDermott 2010). Despite that, she was reported to have been openly flirting with the gang and even danced with Ned.

Her inn was completely destroyed by the fire that the police lit and she and her children were left homeless. To add insult to injury, on the day that Ned Kelly was hanged a warrant for her arrest was issued on the charge of receiving, harbouring and maintaining outlaws. She was imprisoned at Beechworth to await trial, then was acquitted by the jury.

At the age of 78 she gave an interview to a reporter, B.W. Cookson, in which the impact of the loss of her inn and children still clearly brought up strong emotions:

> *Let me begin by saying that I was between two fires there. The police were suspicious of me, because they believed I assisted the outlaws. I did not. The Kellys hated me because they believed I gave the police information about them. I got nothing but abuse and mischief from both sides. And I never had anything to do with either. That is the truth.*
>
> *I well remember Kelly coming to my place that dreadful night. It was raining, and very wet. He took me and my dear little girl away, and locked my two little boys up in a room by themselves. He made me turn the key – said he would shoot me if I refused to do everything that he told me. I begged him to lock myself and my daughter in my own room, but he wouldn't …*
>
> *It was a terrible day. And when the police came and started firing bullets into the house – it was full of people – it was awful. Brave police! They lay in the gullies, and behind the trees, and shot bullets at the house, knowing that it was full of people. My poor innocent little children suffered most. My little boy was shot …*
>
> *My brave little girl was shot, too – shot with a big rifle bullet that had gone through half the house first, or it would have killed her. The bullets were coming all through the house, tearing through the walls, smashing everything, and … Oh, my poor, innocent children! I shall never forget them.*
>
> *My poor little boy was mortally hurt. But no one had mercy. The police kept on shooting, and no one knew who would be the next to fall. The bullets were doing the outlaws no harm at all. They were only hurting us …*
>
> *When my dear little boy was hit he stood up, looked round, and then fell down. 'Oh, God,' he cried, in such a piteous voice, 'Mother, dear mother, I'm shot!' …*

I could not get to the poor child for some time. He was lying on the floor, bleeding from a great bullet wound in his little back ... The murdering police! They had killed him! ... When I got to him I turned him over. He was all blood ... I found the hole ... It was terrible ... His life-blood was pouring out of it, and his poor little white face was turned up, the eyes looking into mine as though imploring help ... Oh, my God, forgive those who did this thing! ... I tore off part of my apron, and tried to stop up the hole in his back with it ... I wanted to go out for help, but Dan Kelly would not let me. 'You can't go,' he said. 'We're turning the prisoners out now.'

My boy died. Died miserably and without help. And my brave little girl, who was wounded herself, never got over it ... She died not long after ... And it was her brother's awful death that killed her! He was such a clever, quiet boy! ... Oh, dear! Oh, dear! (Adelaide Advertiser, 1911).

Ann Jones spent many of her years campaigning doggedly for financial restitution for the destruction of her hotel and compensation for the death of her two children. She was eventually awarded £265, far less than the £5000 she had sought. She was also never able to obtain a publican's licence again, despite many applications (Wright and McDermott 2010).

One of her sons, Terry Jones (also referred to as Jerry) moved to Western Australia and became a policeman, as did Ned's half-brother John King (see p. 27).

Craig Cormick

roof addition to the north. Costing £200, the building was constructed of weatherboards painted white. It had an iron roof and at the time was claimed to be better made than McDonnell's Hotel on the opposite side of the rail track. Photographic evidence taken at the time of the siege shows the roof in pristine condition; it also appears to be painted white (see Fig. 18.1).

The frame of the inn was sawn timber forming the wall studs, wall plates, bearers, joists, rafters, battens and ceiling joists. These were joined with hand-made square-head nails. Flooring would have been abutting pine boards about 1.25 inches (3 cm) thick.

Vertical frame posts would have been set into post holes but the floor bearers and wall base plates may have sat on floor joists, which in turn sat directly on the ground.

The doors were panelled and possibly imported from the US (a common source of finished timber products such as doors at the time) and the parlour door was part-glazed with coloured glass panels (most likely paper transfer rather than leadlight).

The chimneys were locally made terracotta-coloured brick. The internal fireplaces were rendered up to a chest-high timber mantle.

The walls would have been lined with stretched hessian, which would have been painted. The ceilings would have been calico. The skirting would have been 10 inch (25 cm) red pine (Californian redwood) and it is likely that the three main public rooms would have had pine dado panelling. The windows would have been double-hung with square glass panels.

Apparently better fitted out, the Glenrowan Inn also had a better bar counter than McDonnell's Hotel, and had floorboards as opposed to earthen floors. Evidence given at the

Fig. 18.1: Ann Jones' residence and the inn from a photo taken during the siege.

Royal Commission following the siege explained that the inn was doing well and it was patronised by working-class people (Vic. Parliament 1881).

The furnishings are thought to have included the following:

- metal-frame beds (two);
- lighting – oil lamps probably made of brass and glass;
- furniture would have been basic period pieces including dining tables, sideboards, parlour tables and bar chairs;
- water basins and large ewer jugs made of enamelled metal and earthenware;
- water butts for carting water from the spring near Old Sydney Road to the north-east;
- various drink containers made of glass, pewter and pottery, such as demijohns, flagons, bottles and glasses, beer and spirit barrels;
- crockery and cutlery, ornaments and miscellaneous fixtures and fittings such as coat hooks;
- personal items.

Analysis of photographic evidence showed that the inn measured approximately 30 feet (9.1 m) wide by 15 feet (4.5 m) to the rear of the main gabled structure with an additional 10 feet (3 m) to the north incorporating the bedrooms. The southern wall, facing the rail reserve, had two doorways – one into the bar and one into the dining room – and three windows, one each for the parlour at the west, the bar in the middle and the dining room to the east. A verandah ran along the southern wall; the verandah roof was supported on six timber posts. The main gabled portion of the inn had brick fireplaces and chimneys at the east and west ends. There were no windows or doorways at the gable ends.

Less is known of the northern wall as there are no photos or drawings of the inn from the north. It is known, however, that a single door behind the bar gave access to a short hallway between the two bedrooms. A rear door at the end of the hallway opened into a

Mrs Jones' Inn

Mrs Margaret Reardon, the wife of a railway plate-layer, was one of the people taken captive by the Kelly gang during the siege at Glenrowan. She described the layout of the Glenrowan Inn in her evidence given to the Royal Commission in 1881. The transcript of her testimony, with both questions and answers, reads:

> 10552 And did the firing commence immediately? – The firing commenced in about fifteen minutes after that.
>
> 10553 Did you see what Constable Bracken did after he told you that? – No, I did not see him after that, for we were in the back room.
>
> 10554 Can you tell us how the rooms were situated? – There were three different rooms – the dining room, the bar room in the middle, and the little parlor, at the Benalla end.
>
> 10555 How many doors in the front? – Two –one from the bar and the other from the dining room.
>
> 10556 How many rooms were there behind? – Two? and a little narrow passage.
>
> 10557 Did the passage go right through the house? – Only from the bar to the back kitchen.
>
> 10558 It was only between the two back rooms? – Yes.
>
> 10559 And in coming through it you entered the bar coming to the front? – Yes.
>
> 10560 Do you remember the time the firing commenced? – Yes.
>
> 10561 Can you describe anything you saw then in the house? – I did not see anything in the house just then, for there were very dark window blinds, and they were all drawn down.
>
> 10562 Was there no light? – No.
>
> 10563 How many windows were there in the front? – There were three windows on the front – one in each room, and one in the bar.
>
> 10564 Was there a door going from the verandah into the dining room? – Yes.
>
> 10565 And one from the verandah into the bar? – Yes.
>
> 10566 No door from the verandah into the little parlor? – No; you entered that from the bar.

> 10567 Was there a skillion room at the back? – There were two.
>
> 10568 And then at the back of those a small yard? – Yes.
>
> 10569 And two or three steps up to the kitchen at the back? – There were no steps; it went straight on up a little incline to the kitchen at the back.

narrow breeze-way between the inn and Ann Jones' residence and allowed access to and from the kitchen – which was in the residence – and the dining room in the inn (see Fig. 18.2).

While McDonnell's Hotel was considered to be the 'Irish' pub and was known to be where Ned, the gang and their sympathisers drank, Ann Jones' inn served the remaining community and, it is thought, strived to provide a slightly more refined environment (see Fig. 18.3).

Why Kelly chose Ann Jones' inn when things went wrong, rather than McDonnell's, is not clear but it has been suggested that Ann Jones was an informer, or at least could not be trusted, and Kelly chose her inn in order to keep an eye on her and her children.

The siege ended with the torching of the inn, which destroyed the building and the residence that sat only 5 feet (1.5 m) to the rear. The brick chimneys survived and up against the western chimney Ann Jones built a modest hut as a temporary home until compensation for her losses was agreed on and paid. However, she had to wait 18 months until after the Royal Commission into the siege had been concluded (see Fig. 18.4).

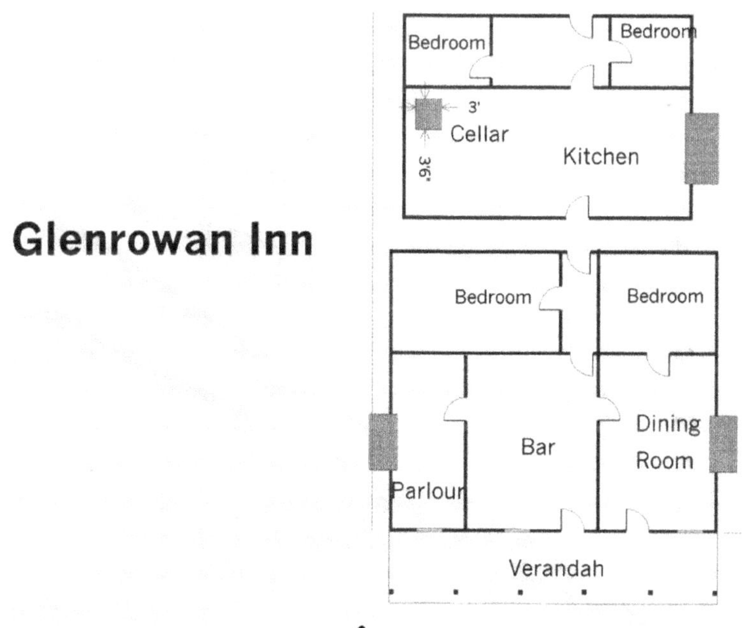

Fig. 18.2: Schematic drawing showing internal divisions of the inn (proportions are approximate).

Fig. 18.3: McDonnell's Hotel just after the siege. Note the coffins for Dan Kelly and Steve Hart on the dray.

With the £265 Ann received in compensation she commissioned another building. It was a large weatherboard house on what was now a corner block. Unable to regain her liquor licence, she leased the building to the Victorian police for £57 a year. The lease agreement described a cottage and stabling on the property.

Very little is known about this building and there is only one photograph of it. However, it appears to have occupied an area slightly larger than the original inn, and extended further to the north. It is also likely that during the time between its construction in 1882 and 1902 when it burnt down, additional outbuildings were constructed to the rear. The location and function of those structures is not recorded.

The police occupied the building until 1895 when a new police house to the east along Siege St was completed. Ann Jones reoccupied the house at that time. Sometime before 1902 a liquor licence was given and the 13-room building became a wine shop and hotel. Ann

Fig. 18.4: A detail of one of Carrington's sketches showing the temporary hut on the inn site.

Fig. 18.5: The second inn or wine shop.

Jones (who remarried and was then known as Ann Smith) moved away and leased the wine shop and hotel business (see Fig. 18.5).

The land then remained vacant, and sometime before World War I a third and final building was constructed on the site. This time the building was brick and built on a floor plan similar to that of the second timber wine shop. It was known as the Café Royal. It was also run as a wine shop (locally known as a wine shanty) and had a common room, dining room, parlour and bedrooms.

In the early 1940s it was renamed the Last Stand Café. Photos show the building with a hipped tin roof and splayed corner entrance and a bull-nosed verandah. Over the years it functioned as a cafe, wine shanty, guest house, confectionery and drink shop and even as a hair salon (see Fig. 18.6).

To the rear were several outbuildings including a euchre room (a card game popular at the time) running from the rear of the building along Beaconsfield Parade, ancillary structures between the main building and the sheds and stabling in the north-east corner of

Fig. 18.6: The third building, as the Last Stand Café.

the lot. The site was cleared in 1976–77 by the present landowner and many of the bricks were salvaged, with the majority of the demolition rubble being removed from the site.

The land has remained vacant since then. The owner planted potatoes in the northern portion of the site for approximately three seasons but the remainder of the site has been untouched, and tourists visiting Glenrowan for many years found only an overgrown lot where the inn had once stood.

Observations of excavation

While it seemed likely that remains of the original inn would have been removed or substantially disturbed by the subsequent occupations, we found surprisingly intact evidence of the foundations of the main gabled section and deeper archaeological elements of the rear rooms. Foundation remains included large post holes forming the outline of the building. Some had the charred stumps of posts surviving. The main gabled section of the inn measured 9 ×4.5 m (approximately – the wine shanty foundations have removed the post holes for this internal wall). The dimensions of the parlour, bar and dining room are unknown and no archaeological evidence was found of internal partitions.

Other artefacts included burnt floor joists that supported the timber floorboards, and the timber chocks that were used to level the wall plates and floor joists. In addition, fallen timbers that had carbonised and iron nails were found in the southern portion of the site.

Machine excavation ceased once the tops of the brick footing were encountered. Archaeologists then excavated in each 'room' of the wine shanty structure and areas immediately to the north and east of the external wine shanty walls. Physical remains of the inn and associated artefacts were discovered surviving between the concrete and brick footings of the wine shanty and the post holes of the second inn.

The stratigraphy (i.e. the particular layers that were developed over time) was generally quite simple. A mixed and disturbed demolition layer overlay the brick foundations and extended across the site. This continued down between the walls and spread out beyond the footprint of the 20th-century building. This dated to 1976–77 when the wine shanty and outbuildings were demolished; it varied in thickness below the level of the brickwork, but was generally 20–80 mm thick.

How timber burns

Timber burns in two ways. In the presence of oxygen, timber will burn away almost entirely to ash, which is fragile and survives in archaeological contexts as a general black or grey layer with no identifiable features of the original timber object. Ash was found in patches across the site, particularly in the south-eastern corner and north-western portion. Timber burnt in the absence of sufficient oxygen carbonises and often retains some of its original form. This happened on the inn site, suggesting that when the superstructure collapsed it smothered part of the burning building, eliminating oxygen and allowing carbonisation to occur.

Adam Ford

The demolition material consisted of mottled and highly mixed material, incorporating soil, clay, mortar, concrete render, crushed brick, paint, plaster, charcoal and a mixture of modern artefacts such as drinks cans and plastic containers as well as early 20th-century building material and the occasional personal item. The majority of artefacts recovered during the project came from this period (see Fig. 18.7).

Below the demolition level the stratigraphy was slightly more complex. Adjacent to the wine shanty foundation walls, and spreading out 300–900 mm from the excavation of the foundation trenches, the remains of the siege-era inn could be found. The deposits were confusing as they varied in colour and composition ranging from light orange or pink – similar to the subsoil – to dark brown and ashy. This layer included fire-affected artefacts from the siege mixed with mortar and render from the construction of the wine shanty (see Fig. 18.8).

It was assumed that the wine shanty foundation trenches were excavated by hand using shovels and the material dug was discarded to either side of the trench. Therefore it was concluded that although this layer included artefacts from the siege that were moved from their original location, it was likely that they were no more than 1 m away from that point and they still had interpretable value.

In some locations directly below the demolition level, physical remains of the inn and remains from the occupation of the inn and siege fire were found.

However, such remains were not consistent across the site. Archaeological deposits from the northern quarter of the inn had been removed when the natural slope was cut away to

Fig. 18.7: Site shot showing rubble and mottled demolition layer.

Fig. 18.8: Site shot showing foundation trench spoil – note different soil colour on right edge of room.

accommodate the wine shanty building and only the lower portions of the foundation post holes survived, cut into the pink granitic sand subsoil.

While the outline of the building could be interpreted from these post holes, nothing remained of the occupation or siege fire deposits. These deposits were, however, found fairly evenly across the remainder of the site, and the best preserved deposits were found in the south-eastern corner of the inn site (see Fig. 18.9).

Once the demolition foundations and deposits were removed the shadow-like foundations of the inn were revealed. These consisted of:

- **Post holes.** Large post holes delineated the outer walls of the main gabled structure and the add-on bedrooms. Post holes of the main structure were on average 900 mm apart (to the centres) and 650–750 mm in diameter. They consisted of irregular holes and post pipe but no post pad. In three instances charred timber posts survived. These features showed where the main supports for the inn's superstructure were positioned. The surviving posts were rough-cut, unshaped logs about 250 mm across. Some would have extended to the top plate of the walls and others would have been stump bearers for the walls. The main gable structure was light-weight and would have had a single-span roof truss, therefore it was not necessary to have mid-span supports and no post holes were found in the internal spaces of the inn (see Figs 18.10, 18.11).
- **Floor joists.** The inn had a floor of sawn floorboards supported above the ground by joists that were in turn sitting directly on the ground. The joists survived as carbonised

Fig. 18.9: The inn site fully excavated.

Fig. 18.10: Post holes of eastern wall of the inn.

Fig. 18.11: Half-section post hole of the south-east corner support. Note the timber still present.

ash deposits sitting within shallow channels. The carbonised features were on average 100 mm across suggesting that the joists were 4 inches across. It is likely that they were 4 ×6 inch (100 × 150 mm) milled timbers. Oriented east–west, the joists were not evenly spaced, varying from 560 mm to over 1300 mm centres.

Most (five out of nine) were, however, spaced around 740 mm apart. The alignment of the joists meant the floorboards ran north–south. At irregular intervals the joists appeared to have been supported on small chocks of wood. It is thought that these were used to maintain the level of the joist across an irregular ground surface. The chocks varied in shape and size and are thought to have been off-cuts. All of these survived in remarkably good condition and do not appear to have been affected by the fire (see Figs 18.12, 18.13, 18.14).

- **Subfloor deposits**. The floorboards would have been 1.25 inch local pine, perhaps Murray pine, and would have been milled locally. The boards would have abutted each other rather than locked together as do modern boards that have tongue-and-groove mouldings. They would have been laid green; as they dried, gaps would have formed between the boards, allowing small objects to fall below the floor. The fire destroyed the floorboards so any objects within the inn fell through to the ground, so it was impossible to tell subfloor deposits from the siege fire deposits. However, it is likely that some of the artefacts recovered would have come to rest below the floor before the siege. One possible example was a penny coin found in the south-east corner of the inn, dating to 1876. Although it was found within the burning layer

Fig. 18.12: Photo showing floor joists and supporting chock.

Fig. 18.13: Site shot of a carbonised floor joist.

Fig. 18.14: Detail of chock in the ground after the carbonised floor joist was removed.

associated with the siege fire it was not heat-affected and may therefore have predated the fire.

- **The siege fire**. The fire that the police lit and that brought the siege to a climatic end destroyed the inn. Photographs taken at the time of the siege show the superstructure of the inn completely failed and it collapsed in on itself. The timber mostly burnt away, leaving the two brick chimneys and the verandah and verandah posts.
  Archaeological evidence of the fire was found across the site and included significant ash and carbonised timber related to the superstructure. As well as the joists and charred posts described above, another piece of carbonised timber was found, presumably where it fell, in the south-east corner of the inn.
  Other evidence of the fire included artefacts clearly affected by heat. A large amount of glass was found that had been deformed and in many instances was completely melted by the fire. Some pieces could still be identified as wine or beer bottles and drinking vessels but most of the glass appeared to have run and frothed or boiled, resulting in irregular shaped and highly aerated lumps of glass. Lead objects were also found, particularly where the rear wall of the bar, dining room and parlour would have been, and the wall dividing the main inn structure and the rear bedrooms.
  The lead had also been deformed and melted by the fire, making identification of the original shapes difficult. While some objects may be melted remains of roof flashing it is thought that most were bullets or parts of bullets, fired into the building by the police during the siege. It is postulated that some bullets penetrated the southern wall of the inn but slowed sufficiently to be halted by the rear wall. These were then

Fig. 18.15: The inn chimneys standing after the collapse of the superstructure.

melted in the fire. Other heat-affected items included tableware ceramics that showed discolouration and crazing of the glaze.

- **The chimneys**. Photos taken after the fire showed the east and west chimneys still standing. However, apart from a small patch of crushed brickwork near where the eastern chimney was approximated, no evidence of either chimney survived. It is likely that the bricks were recovered before the construction of the second inn building. As they were only four years old the bricks would have still been in reasonable condition and it is possible they were incorporated into the new chimneys (see Fig. 18.15).

The gun battle

The siege at Glenrowan is defined by the gun battle that took place between the Kelly gang and the colonial police. First-hand accounts talk about significant force used by the police, amounting to thousands of rounds fired into the inn building. It was hoped that remains of the battle, such as cartridges and projectiles, would be found but that hope was tempered with the knowledge that the site had been salvaged for souvenirs immediately after the siege and had been targeted by metal detectors over the years. It was therefore surprising that projectiles, cartridges and percussion caps were all found, particularly concentrated along the wall separating the front three rooms (the parlour, bar and dining room) and the rear north-west bedroom.

These were highly significant finds for two reasons. First, these artefacts represented definite evidence of the famous gun battle. Second, for the first time it was possible to map and visualise the locations and movements of the members of the Kelly gang during the last hours of their lives (or freedom, in Ned Kelly's case).

Fact or fiction: Dan Kelly and Steve Hart escaped death at Glenrowan

There is very little evidence for this persistent myth that both Dan Kelly and Steve Hart escaped being killed at Glenrowan in 1880 and moved to Queensland, although several people have claimed to be Dan Kelly. The story usually goes that both Steve Hart and Dan hid in a beer cellar at Jones' Inn when it was set alight by police, or crawled out of the building while it was burning and escaped into the bush.

The first contender was an Ipswich vagrant, James Ryan, who walked into the office of the *Brisbane Truth* newspaper in 1933 and claimed to be Dan Kelly. However, the facts of his life as he gave them to the journalist were full of inaccuracies, including his parents' names. Nevertheless, in 1934 he was reported to have appeared at the Brisbane Exhibition Grounds in sideshow alley as Dan Kelly, where people paid to question him about his life.

According to his story he escaped the fire at Jones' Inn, crawled into the scrub with agonising burns and watched his brother Ned's famous last stand against police. Supporting his claim, he was said to have severe burns on his back and the initials DK branded on his buttocks.

Another contender for being Dan Kelly was a Jack Day (or Jack O'Day), who is buried near Mount Isa. It was reported that both Dan and Steve Hart escaped to Queensland and went by the names of Jack Day and Fred Layton.

A third contender for being Dan Kelly was another Queenslander, named Charles Tindall. He was also said to have burn scars on his body. Tindall ran a dairy farm at Oakey, near Toowoomba, married and had four children.

An even more unlikely story was reported by journalist William Bede Melville who, while in South Africa reporting on the Boer War, claimed to have been introduced to two men who identified themselves as Dan Kelly and Steve Hart. A report in the *Moolong Argus* of 1914, upon Melville's death, stated:

> *When the Australian contingents were being sent to South Africa Mr Melville took ship to Capetown, and while there did some vivid work for the London Daily Mail, including the sensational story of the discovery of Dan Kelly and Steve Hart, the Victorian bushrangers, alive and well, and performing prodigies of valour with the British troops. It was explained that instead of being shot in the last stand of the Kelly Gang at Glenrowan these two bushrangers escaped before the hotel was burned down, and had since lived under false names.*

The bodies of the two men more widely believed to be Dan Kelly and Steve Hart were burnt beyond recognition after the police set fire to Jones' Inn but a witness at the scene, Thomas Carrington, a journalist and artist for *Australasian Sketcher*, wrote:

> *The bodies of Dan Kelly and Steve Hart could now be plainly seen amongst the flames, lying nearly at right angles to each other, their arms drawn up and their knees bent, and I feel perfectly certain that they were dead long before the house was fired.*

Even more unlikely is the story that Dan Kelly was executed at the Old Melbourne Gaol in 1880 in Ned Kelly's place. The theory has it that the authorities recognised they had the wrong man, but kept it hushed up.

Steve Hart, photographed by W.E. Barnes, c. 1878. State Library of Victoria.

Dan Kelly, photographed by James E. Bray, c. 1877. State Library of Victoria.

A burned body recovered from the inn at Glenrowan, believed to be either Steve Hart or Dan Kelly, photographed by John Bray, 29 June 1880. State Library of Victoria.

Craig Cormick

As mentioned previously, lead objects had been found along the southern face of the internal wall line, suggesting that it afforded some protection against the raking fire that was penetrating the front wall. Excavations on the northern side of the wall line found a concentration of cartridges and percussion caps. The location of these cartridges and caps, found where the rear bedroom been, suggested that they were from the Kelly gang's own weapons and that members were reloading in an area of relative safety (perhaps the bar counter afforded protection as well, as it would have stood between the corner of the bedroom and the main police lines). The gang members would have then gone back into the lethal zone of the main bar to engage the police through the windows and doorways.

Comparative analysis of strike marks on one percussion cap with Ned Kelly's revolving carbine, currently in private hands, showed a match. Not only were the cartridges from Dan Kelly's or Steve Hart's or Joe Byrne's weapons, but at least some of the percussion caps had been discarded by Ned Kelly from his assorted archaic weapons, during the early stages of the gun fight (see Figs 18.16, 18.17, 18.18, 18.19).

Human remains

Ned Kelly had been captured some way to the east of the inn and Joe Byrne's body had been recovered by the Catholic priest Father Gibney, who ran into the burning inn to do so, but the bodies of Steve Hart and Dan Kelly were consumed in the fire. Father Gibney noted, and other eyewitnesses confirmed, that Dan and Steve were lying in the western bedroom already dead, either by their own hand or from wounds sustained during the battle. Photographs of one of the bodies recovered after the fire showed that the extremities (the arms and legs) had burnt away; the project team thought it possible that fragments of bone may have survived in the ash contexts in that location of the inn.

All ash deposits in the north-west portion of the site were therefore sieved. A small number of bone fragments that showed evidence of significant heat were indeed recovered. This bone was deposited with the Victorian Institute of Forensic Medicine at the State Coroner's Office in Melbourne for analysis. Unfortunately the analysis was inconclusive, due to the small size of the bone fragments and few samples. The largest fragment found appeared, from visual evaluation, to be from an animal (see Fig. 18.20).

The invisible contexts and the ghost contexts (discussed previously) were indicated by obvious absences from the site. No weapons, belt buckles, armour or powder canisters were found and only comparatively small amounts of ceramic were recovered. Also, only a small number of cartridges were recovered, considering the length of the gun battle. No furniture or fixtures or fittings were recovered at all. It is assumed that obvious artefacts and items that would be readily identifiable as coming from the siege and therefore being of value, such as cartridges, pewter mugs, plates, weapons etc., would have been recovered from the site immediately after the siege.

Archaeological evidence seemed to show that metal detectors had not significantly affected the integrity of the site although four pits, mostly dug in the western part of the site, were identified as recent robber pits, where people had dug looking for artefacts.

Fig. 18.16: Lead objects thought to be melted projectiles.

Fig. 18.17: Another example of the melted lead objects.

Fig. 18.18: One of several cartridges recovered from the site.

Fig. 18.19: One of several percussion caps found.

Fig. 18.20: Location of possible human bone remains.

Reflecting on the past

Historical accounts are often all that is available to learn about what happened in the past. In general, histories are incomplete but, more importantly, they can be inaccurate – they are often biased towards the victor, or even just the author of the account.

The Kelly story is more complicated than most because the story is greater than the facts that provided the initial historical account. And the story has been personalised and exaggerated and manipulated so that there are accounts that, with great authority, pronounce the Kellys, particularly Ned Kelly, as heroes or as common murderers. Uniquely in Australia's history, everyone seems to have an opinion on this event.

This situation, however, provided an opportunity and a challenge to the archaeological project team. The opportunity was to add fact and substance to the legend, while the challenge was to conduct the project as impartially and sensitively as possible.

Fortunately, we succeeded in both. The excavation provided physical evidence of the battle and brought the story back to the human scale, which is often lost when an event becomes legendary. The dig uncovered the carbonised shadow of the small timber inn, displaying rooms where it is difficult to imagine 62 people trying to shelter as bullets ripped through the walls. It located the spot where Joe Byrne died and where Dan Kelly's and Steve

Hart's bodies were destroyed by the fire. It showed the ferocity of the fire, from the carbonised floor joists to the melted lead projectiles and the bubbled and deformed glass.

Tiny percussion caps and cartridges discharged from the Kelly gang's weapons were discovered and handled for the first time since Ned and the others threw them to the ground while reloading in their final hours.

For the team working on the dig, the site was now stripped of the legend and romanticism. We could better visualise the fear and panic that the gang and members of the public would have felt in the dark of a tiny-roomed simple bush pub, as many hundreds of bullets ripped into its walls. For the Kelly gang this wasn't a heroic stand but a fight for their lives after a plan gone wrong. And it led to the violent deaths of three people and the fatal wounding of two others – including Ann Jones' own 13-year-old son.

References

The Kelly gang: new light on an old tragedy. *Adelaide Advertiser*, 23 September 1911.
Wright C, McDermott A, ((2010)) Ned's women: a fractured love story. *Meanjin* **69**(2), 113–121.

Chapter 19
The police perspective

Elizabeth Marsden

Victorian police are predominantly presented as villains in most popular tellings of the Kelly story. Examination of police archives, however, reveals a much more complex story, and provides insights into the difficulties faced by the men bound to upholding the laws of the day.

> … a parcel of big ugly fat-necked wombat headed big bellied magpie legged narrow hipped splaw-footed sons of Irish Bailiffs or English landlords which is better known as Officers of Justice or Victorian Police (Kelly 1879).

Victorian police have not fared well in the history of the Kelly gang. Indeed, Ned Kelly's comical, often venomous depictions of the men sent out to capture him continue to be used to support the popular image of an inherently corrupt force. Unlike Ned, police have been rarely given the opportunity to explain or justify their actions. This occurs despite both sides being fundamentally linked and equally important to the understanding of what is one of the most complicated and controversial chapters in Australian history. While most Australians identify Ned Kelly as a national folk hero, few understand the real events in which he was involved. Over time historical fact and myth have become blurred, with the story now firmly dominated by one voice who, it must be admitted, was a convicted criminal. Police archives challenge this tradition, provide different perspectives and voices, and help to contextualise and broaden our understanding of the history, by indicating the difficult social dynamic that existed in Victoria's north-east at the time; part terrorised, part supportive of the gang.

Predominantly held by the Public Record Office of Victoria and the Victoria Police Museum, police archival records are highly standardised and thorough, and often lack the emotion expressed in other written accounts. Nonetheless, the information recorded in the various police watchhouse books, occurrence books, service records, reports, letters, memoranda, telegrams and even photographs offers intimate details of the lives of individuals, their education, social condition, beliefs, physical traits and even mannerisms. They also provide valuable insights into early police investigation methods, communications and resources available to them, as well as the rationale behind some of the decisions made. With the hunt for the gang lasting almost two years, the trail of documentation generated constitutes critical data in tracing the Kelly story and its impact upon the wider community.

Seeing the police as individuals

Presented as what researchers term a 'homogenous construct' (a single group) police from the Kelly period have traditionally been denied the level of individuality afforded to the gang members, their families or even their extended associates, and tend to be simply described collectively as the police. The exceptions tend to be only the more dubious members of the police – despite the existence of hundreds of commendable police service records from the Kelly period. Police collections help to address this imbalance. They do not show an impeccable force, but rather a poorly trained, under-equipped band of men, some of higher probity than others. However, the image of a villainous police force has become crucial to the popular narrative that justifies the gang's violent actions.

Founded in 1853, less than two years after the separation of the colony of Victoria from New South Wales, Victoria Police have been present at every significant milestone in Victorian history, including the gold rush, the Eureka Stockade and the infamous Kelly gang outbreak of 1878–80. While its administration and discipline procedures were largely drawn from the London Metropolitan Police (a civil policing body), operationally Victoria Police was modelled on the Royal Irish Constabulary (RIC) who were regarded as a militaristic, repressive political tool, well accustomed to the use of informants and secret surveillance. The influence of Irish constabulary methodologies and its reputation on Victoria Police was profound and reinforced from 1870 to the 1880s, when all new recruits were required to first serve in the Victorian Artillery Corps.

The force was overwhelmingly Irish in representation: historian Robert Haldane's analysis of police oath sheets from 1874 indicated that 82% of Victorian officers were Irish, with 46% having served in the RIC. He argued that while the political climate in Ireland differed considerably from that in Victoria at the time, it is reasonable to assume that many in the colony accepted certain RIC methods and ethics as routine (Haldane 1986). Whether this was true for recruits entirely new to policing is, however, unknown.

A climate of fear

The popular rendering of Ned Kelly as a social and political rebel is often explained as 'social banditry', whereby rural bandits emerge from regional peasant societies during periods of economic hardship and political turmoil as champions of the poor and oppressed. The real exploits of genuinely famous bandits often matter less than the mythology surrounding them (Hobsbawm 2000). The story of the Kelly Outbreak is no different.

Ned by his own admission was an accomplished horse and cattle thief, and may even have practised this on a professional scale (Morrissey 1987). While the widespread stock theft experienced in Victoria's north-east during the Kelly period is often presented as an indicator of rural poverty, Morrissey suggested such activity was specific to minority groups, such as the 'Greta Mob'. Chiefly drawn from the sons of local selectors, including Ned Kelly and his friends, many members of the mob would later be identified as Kelly gang sympathisers. With nearly 40 members around the Greta and Glenrowan area, the Greta mob was notorious for horse and cattle theft and general anti-social behaviour: 56% of members, including Ned Kelly himself, had criminal convictions before the Kelly Outbreak. Targeting rich and poor alike, the mob moved stock overland through an elaborate system of back-country safe-houses manned by extended family and criminal cooperatives. Plough horses were the most regularly stolen.

While many charges brought against the Greta mob were later dismissed for lack of evidence, they were not unfounded. Journalists of the time spoke openly of a climate of fear in the area, with many in the community wary of speaking out against the group for fear of reprisal:

> … for some considerable time the respectable part of society around Mansfield has been living in a state of terrorism, submitting to be robbed right and left and afraid to complain for want of sufficient protection (*Age* 1878).

This situation naturally hampered police investigations, as described by Assistant Commissioner Nicholson during the 1881 Longmore Royal Commission following the Kelly Outbreak:

> We could get little or no assistance from the inhabitants, and the people were all through the country in such a state of terror. Civility was shown us in every town in the district, but no information given. The people seemed to be more afraid of the gang than confident in the police (Royal Commission 1881).

The charges outlined in police ledgers, such as the Greta watchhouse book, more accurately indicate the level of criminal activity in the area than do conviction rates alone. One of hundreds of such books held by the Victoria Police Museum, the Greta watchhouse book schedules charges brought against several Kelly family members and known associates at the time of their initial apprehension between 1870 and 1873. Arrests and crimes documented include Edward Kelly, assault and horse stealing; Dan and James Kelly, illegally using a horse; Ellen Kelly, stealing a saddle; John Lloyd, cattle stealing; as well as James Quinn, Ned's uncle, and Kelly associates Isaiah Wright and William Williamson for grievous bodily harm (see Figs 19.1–19.4).

These entries provide insight into the character and behaviours of individuals in ways that contradict the popular imagination. Yet despite the evidence of the Kellys' extensive involvement in organised and petty crime long before the actual outbreak, the police are routinely presented as the antagonists, specifically Constable Alexander Fitzpatrick (see Fig. 19.5).

Constable Fitzpatrick

Undoubtedly of unreliable character, Constable Fitzpatrick attended the Kelly home on 15 April 1878, intoxicated and without an official warrant, to arrest Dan Kelly on a charge of horse stealing. Documented eyewitness accounts of the violent altercation that followed between the household and the officer are contradictory and confusing, with all involved having motive to manipulate the facts. Marking the beginning of life on the run for the Kelly brothers and the imprisonment of three people, including Ned Kelly's mother Ellen, the incident was further tainted by Fitzpatrick's inability to follow correct police procedure. He would remain a thorn in the side of Victoria Police throughout his limited three-year career, before being unceremoniously dismissed in 1880 for 'general misconduct'.

Unjustifiably presented as an example of all officers of his day, Fitzpatrick's record of conduct and service demonstrates that his persistent incompetent and corrupt behaviour was not considered normal and nor was it tolerated. The habitual emphasis placed on a small number of

287

No. 10926 Name Kelly, Edward

Height ... 5 ft 10	Sentence.	Three Years H.L. / Two & a half years [illegible]	Death
Weight ... 11 st 4			
Complexion ... Sallow			
Hair ... Dk Brown			
Eyes ... Hazel			
Nose ... Medium	Date of Conviction.	2. 8. 71	29 October 1880
Mouth ... Medium			
Chin ... Medium			
Eyebrows ... Dk Brown	Offence.	Receiving a stolen horse	Murder
Visage ... Broad			
Forehead ... Low			
Date of Birth ... 1856			
Native place ... Victoria	Where and before whom tried.	Beechworth Gen. Sess. J. [illegible] [illegible]	Melbourne Central Crim. Court Sir Redmond Barry
Trade ... Laborer			
Religion ... R. Catholic			
Read or Write ... Both			

Scar top of head, Two scars crown of do, Scar front of head, eyebrows meeting, Two natural marks between shoulder blades, Two freckles lower left arm, Scar back of left thumb, Scar back of right hand, Three scars left thumb

Single, Mother Ellen Kelly living at Greta, or [illegible] river, Uncle James Kelly a prisoner at Pentridge & another James Quinn now at Beechworth awaiting trial and John Lloyd an uncle a prisoner at Pentridge. 3 [illegible] same day at Wangaratta Pol. Ct. Vag. 3 Mo. [illegible] in default of bail – Bail found

	When received.	Offences, Sentences, &c.	Extensions by— Visiting Justice M. D.	Superintendent M. D.
		Brothers James No. 10861 Daniel No. 10991		
Beechworth Gaol		21. 5. 73 [illegible]	5	7
Pentridge	19. 2. 73			
[illegible]	25. 6. 73	[illegible]		
Battery	25. 9. 73	[illegible]		
Melbourne Gaol	29. 6. 80	11 November 1880. Executed at 10 am.		

Half-yearly Report.

Date.	Superintendent's.	Favorable.	Unfavorable.	Days absent, &c.	Attendance at School as Monitor.

Fig. 19.1: Ned Kelly's police record. National Trust Australia (Victoria).

Fig. 19.2: Ned Kelly. Victoria Police Museum.

Fig. 19.3: Dan Kelly. Victoria Police Museum.

Fig. 19.4: Joe Byrne. Victoria Police Museum.

Fig. 19.5: Constable Fitzpatrick, who accused Ned Kelly of attempted murder, and was later dismissed from the police force. Victoria Police Museum.

bad apples such as Fitzpatrick denies the evidence that most police were considered professional by their contemporaries, doing the best they could, often in very difficult circumstances.

Isolation, fatigue and general overwork

Poorly paid and largely drawn from the working classes, Victorian police occupied the lowest strata of the public service and endured poor social status. Low morale was rife. Unable to vote, marry without permission or go on holiday, police were required to wear their uniforms in public at all times, even when not on duty. Often living in rudimentary accommodation, with the potential of being transferred at short notice, it was often difficult for men to form meaningful bonds with the communities in which they served. In addition, rural officers struggled with isolation, fatigue and general overwork, while police reward systems, naively introduced to encourage diligence, only provided the temptation for perjury and corruption. Also, poor discipline and drunkenness persisted.

Considered a chronic problem of the working classes, police drunkenness was for the most part tolerated by communities, and has been described as an example of the public 'getting the force they deserved' (Haldane 1986). For, while Victorians could have demanded educated recruits and paid them accordingly, the working classes could be employed at less cost and were regarded as more suited to the onerous nature of police work.

In contrast to their modern counterparts, police during the Kelly Outbreak were impeded by a general lack of basic training and resources, as well as ridiculously restrictive regulations aimed at saving the government money. Wishful notions of policemen being born, not made, hampered professional development. Mounted police, such as those sent out in search of the gang, were generally armed only with a sword and a revolver and even then received only sword training (McIntyre 1900). They were issued with only 12 rounds of ammunition, for which they had to remain accountable at all times, and had to bear the cost of any practice firing, being charged at a rate of 6d per bullet, equivalent to the cost of a loaf of bread.

Despite the difficulties with which many men worked, Victorian police during the 1870s received a starting wage of just 6s 6d per day, equivalent to that received by general labourers. Consequently, there was little incentive for men to train themselves, and it was not uncommon for men to retire from the force without ever having fired their weapon (Haldane 1986). In addition, resources were often unequally distributed – correspondence files show that while some stations were amply supplied, others struggled to find sufficient horses, revolvers and ammunition. For example, just two days after the Stringybark police murders Detective Ward sent an urgent telegram to the police headquarters seeking aid:

> No webleys ammunition in store here please send one hundred rounds also a few revolvers (Ward 1878).

Indeed ammunition shortages during the Kelly Outbreak became so dire that all practice firing was banned (Circular memo 1879). This was in contrast to the Kellys' resources; they were reported to be making their own bullets, and were well known for their marksmanship and horsemanship skills.

Unsurprisingly, police shortcomings did not go unnoticed. Correspondence files from the period contain numerous letters from members of the public voicing their concerns.

> … what makes me write this is that -7- troopers are told to endeavour to capture the bushrangers … I consider it is sending men to die for nothing … they are badly armed … and besides are short of ammunition (Peppin 1878).

Representing a diverse social demographic, the letters indicate the impact of the outbreak upon the wider community. Crimean war veterans, doctors, butchers, retirees, miners and even psychics wrote to the Chief Commissioner to volunteer in the search for the gang or to offer suggestions of ways to apprehend them. While the use of bloodhounds, native trackers and guerrilla warfare-type tactics may not have seemed out of the question, other suggestions were more outlandish, such as deliberately starting bushfires, blasting the bushrangers out with dynamite or tempting them out with circus tricks, as was put to Captain Standish:

> … a circus in one of the towns bordering their haunts would bring them out, we all know the love of feats of horsemanship bred in those wretches … they could never resist this (Smith 1879).

While ordinary people sought to play a part in history, many of the institutionalised weaknesses experienced by police would remain unchanged until well after the outbreak. The Longmore Royal Commission later highlighted police deficiencies exhibited during the outbreak, concluding with a list of 36 recommendations for reform, including the introduction of instructional classes and promotional exams, new recruitment procedures and the development of a code of practice for police.

Irish-born police versus native sons

With the overwhelming majority of police having emigrated from industrialised Ireland, the Australian-born gang with their intimate local knowledge and extensive bush skills held considerable advantages over the police sent out to pursue them (see Figs 19.6–19.9). The thick and mountainous bushland of Victoria's remote north-east was quite a foreign landscape to most, with just 80 officers stationed in the remote district of over 28 000 km^2.

Individual correspondence files and reports from the period provide insight into how individuals dealt with and responded to these difficulties, and at times reflect an underlying sense of personal desperation. In the days following the Stringybark Creek police murders, Sub-Inspector Pewtress wrote to his superiors seeking relief:

> I wish you would send here a Sub-Officer who understands bush life and take charge of the men, as I fear in doing so myself I occupy a false position, inasmuch as after a days riding I am thoroughly knocked up, and I know nothing whatever of bush life [and] therefore unable to guide men as to the course they should pursue and might through ignorance lead them into danger [and] perhaps death (Pewtress 1878).

Fig. 19.6: Constable Thomas McIntyre. Victoria Police Museum.

Fig. 19.7: Sergeant Kennedy. Victoria Police Museum.

Fig. 19.8: Constable Scanlan. Victoria Police Museum.

Fig. 19.9: Constable Lonigan. Victoria Police Museum.

Pewtress was by no means an incompetent policeman (see Fig. 19.10). A well respected career officer, he was quite famous in his day for being one of the original 'London Fifty' recruited from the London Metropolitan Police in 1853 to help mould the Victorian service into a well structured and effective force. His predominantly metropolitan experience was his only disadvantage, since it did not prepare him for rugged bush work. His memorandum, written the day he returned from recovering the bodies of the police murdered at Stringybark Creek, is clearly a heartfelt confession of his inexperience of working in the foreign Australian landscape.

The Thomas McIntyre story

Emotion and personal vulnerability are rarely expressed in conventional police operational documentation and tend to be reserved to the handful of personal accounts and memorabilia collections from the police perspective, such as the Thomas McIntyre collection held by the Victoria Police Museum. Witness to the murder of two of his colleagues by the Kelly gang at Stringybark Creek on 26 October 1878, McIntyre's testimony would lead to Ned Kelly's murder conviction and execution.

The critical juncture in the Kelly history, signifying the official beginning of the gang, Stringybark Creek represents a significant departure from the traditional highwayman tradition and in many ways contradicts the popular Kelly myth. Possibly for this reason, it remains one of the most neglected aspects of the popular rendering of the story.

McIntyre's collection, consisting of an unpublished memoir, three scrapbooks of newspaper cuttings and a hand-drawn map of the Stringybark Creek ambush site, provides a

Fig. 19.10: Sub-Inspector Pewtress. Victoria Police Museum.

complete narrative of one man's experiences and understanding of events and provides vivid personal insight into the Kelly story:

> Ned Kelly's eyes have been variously described as grey, dark, and brown. I described them as hazel with a green streak through them, so at least they seemed to me when he held the revolver close to my chest (McIntyre 1900).

Articulate and surprisingly balanced in his analysis of events, McIntyre's memoir has as much validity in understanding the Kelly story as the words of Ned Kelly himself. Combined with Ned Kelly's Cameron and Jerilderie letters, we have two aspects of the same story, one from either side of the law. In addition, McIntyre's map, rediscovered in 2009, is the most detailed extant plan of the ambush site. It provides accurate geographic data on the police camp location as well as positions of the gang and police members, from the time of the initial ambush to McIntyre's escape (see Fig. 19.11).

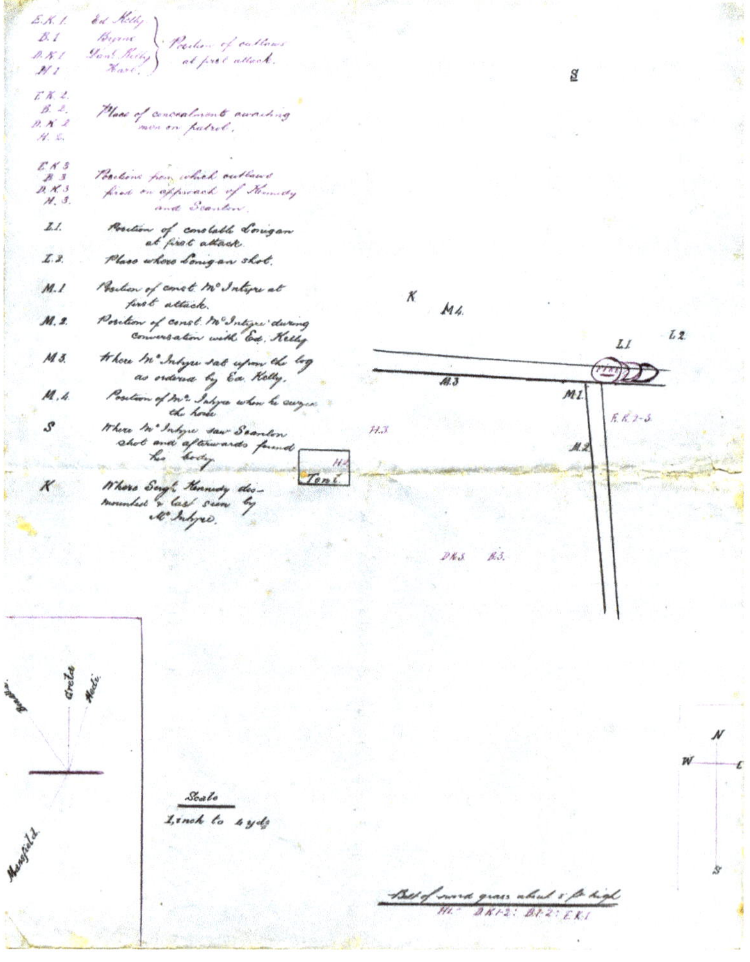

Fig. 19.11: The McIntyre map of Stringybark Creek ambush site. Victoria Police Museum.

McIntyre has been presented by some as a perjurer who altered his testimony to secure Ned Kelly's conviction. The slight variations between his numerous reports and statements concerning Stringybark Creek have permitted a flexibility of interpretation, fuelling the speculation that Kelly was persecuted by police. In the absence of court transcripts from the 1880s and accepting the failure of Kelly's own lawyers to satisfactorily reconcile the constable's statements for a modern audience, McIntyre's collection has an important role to play in disputing the tradition that Kelly was persecuted by police, and potentially vindicating one of the story's most debated characters.

An examination of his materials reveals a meticulous collation of reports and statements that suggest a man obsessed with documenting the events he witnessed, with as much accuracy as he could, not someone intent on obscuring the truth.

Contesting the past

The story of Ned Kelly evokes some of the strongest, most passionate opinions within our culture. Indeed the degree of myth that has gathered around this iconic Australian story indicates its importance to the national imagination. The issue now is that, with so many people identifying with the imagined story, any attempts to question its authenticity or re-evaluate the past can be passionately contested, stagnating historical analysis. As archival collections are digitised and made publicly accessible online, everyone, not just historians, has the opportunity to form their own interpretation based upon how they view the evidence, opening up the story to more varied debate than ever before.

References

Age, 2 November 1878. Cited in T. McIntyre scrapbook, VPM3806, pp. 9–10.

Circular memo (1879) 5/79, Police Department, Superintendent's Office, Benalla, 24 March. Victoria Police Museum, VPM6123.301.

Flood (1878) Report relative to the bullet mould found at Mrs. Kelly's Hut, Greta Station, 14 May. Victoria Police Museum, VPM6123.62.

Haldane R (1986) *The People's Force*. Melbourne University Press, Melbourne.

Hobsbawm E (2000) *Bandits*. Weidenfeld & Nicolson, Great Britain.

Kelly N (1879) *The Jerilderie Letter*. Republished 2001.Text Publishing, Melbourne.

McIntyre T (1900) A true narrative of the Kelly gang. Unpublished manuscript, Victoria Police Museum Collection.

Morrissey DJ (1987) Selectors, squatters and stock thieves: a social history of the Kelly gang. PhD thesis. Department of History, La Trobe University.

Morrissey D (1995) Ned Kelly and horse and cattle stealing. *Victorian Historical Journal* **66**(1), 29–48.

Peppin AB(1878) Letter to Captain Standish, 31 October. Victoria Police Museum, VPM6123.138.

Pewtress H (1878) Memorandum, 29 October. Victoria Police Museum, VPM6123.221.

Smith JJ (1879) Letter to Captain Standish, 25 February. Victoria Police Museum, VPM6123.258.

Victorian Parliament (1881) Second Progress Report of the Royal Commission of Enquiry into the Circumstances of the Kelly Outbreak, the Present State of the police Force etc. J.Ferres, Government Printer, Melbourne.

Ward D (1878) Telegram to the Chief Commissioner of Police, 28 October. Victoria Police Museum, VPM6123.84.

Chapter 20
Ned Kelly's inquisition

Iain West

If present-day coronial law, under the Coroner's Act of 2008, had been applied in 1880, would the coroner's inquest into Kelly's death be managed differently?

To answer this question we first need to look at the inquest that was held into his death in 1880.

The inquisition to inquire into the manner of the death of Edward Kelly was held at the Melbourne Gaol on behalf of Queen Victoria, on the morning of 11 November 1880. The presiding coroner was Dr Richard Youl, who had been appointed City Coroner in August 1854. He sat with a jury of 12 'good and lawful men', as mandated by the Coroners Statute of 1865, summoned from the residents of the colony of Victoria.

Edward Kelly was hanged within the gaol at 10 a.m. that same day, and the place of death gave Coroner Youl jurisdiction to inquire as to the manner of his death. The body was viewed and a jury sworn and charged to inquire when, where, how and by what means Edward Kelly came to his death.

Three witnesses gave evidence at the inquest. The Gaol Governor, Mr Castieau, told the jury that he was present in the Supreme Court of the colony when Edward Kelly was sentenced to be hanged. He further stated that he was present when the execution 'was carried into effect under the provisions of the Act for the Private Execution of Criminals' (governor's signed deposition). The Deputy Sheriff for the colony gave similar evidence and produced the governor's warrant for the execution of the deceased. The jury then heard from the gaol surgeon, Dr Shields, who stated that he was present when Edward Kelly was judicially hanged and on later examination of the deceased found 'that he was quite dead' (surgeon's signed deposition).

Dr Youl recorded the jury verdict and the 12 jurors signed the inquisition finding:

> In the Melbourne Gaol on the eleventh of November current the deceased Edward Kelly was judicially hanged. The jury found that the provisions of the Act for the Private Execution of Criminals was properly carried out (Inquisition finding).

How might things be different today?

It is clear that an inquest would have to be held, due to Kelly's status as a person placed in custody immediately before his death, as outlined in the *Coroner's Act 2008*. However, as an objective of the current legislation is the avoidance of unnecessary duplication of inquiries

Fig. 20.1: Ned Kelly in court. *Australasian Sketcher*, 6 November 1880. State Library of Victoria.

and investigations, the coroner could proceed on hearing a summary of the circumstances surrounding the death, given on oath by the investigating police member.

This could be sourced from the earlier criminal proceedings in the Supreme Court (see Fig. 20.1), together with information obtained from the examining doctor and witnesses present at the hanging (were hanging still allowed).

The holding of the inquest under current legislation could, however, give rise to the appearance of an applicant seeking 'interested party' status at the inquest, either in person or with legal representation. A family member would usually fall into this category. And despite this being a judicial hanging, the family could question, for example, whether the hanging was properly carried out and request relevant witnesses be called.

The appointed hangman in 1880 was Mr Elijah Upjohn, then a convict under sentence. His first and only hanging was that of Kelly; for doing so, he received the sum of five pounds (Finn 1888). A judicial hanging required the neck to be broken. It was not uncommon, however, for hangings to be botched, with the drop being too short, leading to a slow death by strangulation, or it being too long and the head being removed. To address such a concern witnesses would need to be called, with the interested party having the rights of cross-examination and of making submissions.

This process could lead to the coroner making recommendations aimed at ensuring compliance with the legislative requirements for judicial hangings. Such recommendations would then require a written response by the authority to whom they were directed.

At the conclusion of the evidence and any submissions, the coroner would make findings of fact as to the identity of the deceased, the cause of death and the circumstances in which the death occurred. A jury would not be sitting to return a verdict regarding these facts, though, as the holding of an inquest into a death with a jury was abolished in 1999 by amendment to the *Coroners Act 1985*.

References

Coroners Act (1985).

Coroner's Act (2008).

Coroners Statute (1865).

Finn E (1888) *The Chronicles of Early Melbourne, 1835 to 1852: Historical Anecdotal and Personal, by "Garryowen".* Fergusson and Mitchell, Melbourne.

VPRS 24/P0, unit 411, Inquest Deposition Files, inquest # 1880/938, Edward Kelly (VA 862 Office of the Registrar-General).

VPRS 24/P0, unit 411, Inquest Deposition Files, inquest # 1880/938, Surgeon's signed deposition.

VPRS 24/P0, unit 411, Inquest Deposition Files, inquest # 1880/938, Governor's signed deposition.

Chapter 21
Edward Kelly: the last legal rites

John Coldrey

The conduct of Ned Kelly's trial made a verdict of acquittal on the ground that he was acting in self defence a virtual impossibility.

On 30 October 1880, after a trial compressed into two days, Edward (Ned) Kelly was found guilty of the murder of Constable Thomas Lonigan at Stringybark Creek. The jury took only 30 min to reach its verdict.

The trial judge, Sir Redmond Barry, sentenced Kelly to be hanged by the neck until dead. He concluded with the customary pompous legal benediction 'and may God have mercy on your soul'.

Kelly responded: 'I will go a little further than that, I will meet you there where I go' *(Argus*, 30 October 1880).

Kelly was hanged at the Melbourne Gaol on 11 November. Justice Barry died in his bed 13 days later. The judge's death is often given the status of a prophecy fulfilled, rather than a felon's grandiose boast of heavenly favour (see p. 204). The fact that the predicted meeting had no time frame, and that the judge had been in ill-health, is conveniently ignored.

But the millions of words and multiplicity of books written about the exploits of Ned Kelly frequently exemplify a myth-making tendency that has enveloped his legacy.

Was he a hero, fighting against state-sanctioned oppression, or a criminal with a contempt for the law and civil society?

One battleground for the competing contentions has been Ned Kelly's trial for murder. Did he receive a fair trial? Should he have been acquitted on the grounds of self-defence?

Any attempt to answer these questions involves an evaluation of the facts surrounding the shooting of Constable Lonigan and an examination of the applicable law.

This task is made difficult by the fact that no trial transcript is in existence and the primary sources of information, the *Age* and *Argus* newspapers, are not always totally consistent and almost certainly constitute an incomplete record of court proceedings.

Events at Stringybark Creek

On 25 October 1878, a police party of four, Constables Thomas McIntyre, Michael Scanlon and Thomas Lonigan and Sergeant Michael Kennedy (the leader), left the Victorian country town of Mansfield on horseback. Their purported mission was to locate and arrest Edward and Daniel Kelly for horse stealing and the attempted murder of a trooper – Alexander Fitzpatrick. Warrants for their arrest had previously been issued.

Fact or fiction: Ned Kelly's courtroom curse killed Judge Redmond Barry

After being sentenced to death at his trial, Ned Kelly famously replied to the judge, Sir Redmond Barry, 'I will see you there where I go.' The judge died 12 days after Ned Kelly was executed. His death certificate listed the cause of death as pneumonia and anthrax, but it is more likely he died from a combination of pneumonia and an untreated carbuncle on his neck that led to septicaemia, or blood poisoning, made worse by diabetes.

The two men, Kelly and Barry, had been antagonists for some time. Redmond Barry had earlier sentenced Ned's mother to three years in prison for the attempted murder of Constable Fitzpatrick in 1878, which was largely seen to have caused the Kelly Outbreak. A decade earlier, in 1868, Barry had sentenced Ned's uncle, Jim Kelly, to death for arson and attempted murder – a sentence which was later reduced to 15 years in prison. Incidentally, he had also presided over the sentencing of many of the rebels from the Eureka Stockade in 1855.

It was Redmond Barry who had Ned Kelly's trial moved from Beechworth to Melbourne, due to fears that sympathisers of the Kelly gang would interfere with a trial there.

According to the *Age* of 30 October 1880, the courtroom discussion between Barry and Kelly after the death sentence was handed down, involved Barry saying: 'May the Lord have mercy on your soul.' And Kelly responding, 'Yes; I will meet you there.'

The *Argus*, however, reported Ned Kelly's reply as: 'I will go a little further than that, and say I will see you there where I go.'

And the *Bendigo Advertiser* stated: 'He betrayed no emotion whatever, but said he had no fear of death, and when the dreadful words announcing his doom were uttered by the judge, coolly bid good-bye to his friends in court, and walked from the dock.'

Ned Kelly was executed on 11 November 1880 and Sir Redmond Barry died on 23 November. According to the Melbourne *Argus* of 24 November:

> *Sir Redmond Barry had been suffering from diabetes for about 10 years, but the state of his health was not such as to occasion alarm to his friends. On his return from his trip to Europe and America a few years ago, it was apparent to his medical adviser that the disease had affected his system. His Honour, however, always took the most hopeful view of things, and was, if anything, slightly indifferent about the state of his health. On Monday, the 15th inst., he was first troubled with the carbuncle on his neck. Sir Redmond was counselled by his medical adviser to at once rest from duty, but he was reluctant to do so, and continued to attend the court until his disease had such a prostrating effect that he was compelled to take rest … Despite the precautions, however, his Honour caught cold through exposure, and congestion of the left lung set in. Dr. Gunst held a consultation with Dr. Teague, and pronounced the case hopeless. The left lung had become greatly congested, and this, together with the exhaustion and wasting away of the system resulting from the previous disease, proved fatal.*

Craig Cormick

Source: A Galbally (1995) *Redmond Barry: An Anglo-Irish Australian*. Melbourne University Press, Melbourne.

The police contingent made camp in a clearing at Stringybark Creek in the Wombat Ranges some 20 miles (32 km) from Mansfield. By the following evening three of the policeman lay dead at the campsite. Each had multiple gunshot wounds.

Only Constable McIntyre survived the carnage, escaping on horseback to bring news of the shootings to the outside world. Stringybark Creek was about to be transformed from obscurity to notoriety in the history of Australia.

Two years later Constable McIntyre was to become the key Crown witness in Kelly's trial.

According to McIntyre's evidence, at about 5 p.m. on 26 October, when Kennedy and Scanlon were patrolling in the bush on horseback, he heard voices calling 'Bail up! Hold up your hands!' Four armed men emerged from the tall spear grass. A man he identified as Ned Kelly had his weapon pointed at McIntyre's chest and McIntyre held out his hands horizontally in compliance with the command. Kelly then turned the rifle in line with the nearby Lonigan and fired. Lonigan fell, calling out 'Oh Christ, I'm shot'. He died shortly thereafter. Kelly and the gang members (Dan Kelly, Steve Hart and Joe Byrne) then hid themselves and awaited the return of Kennedy and Scanlon (*Argus*, 29 October 1880).

McIntyre remained under surveillance. His designated task was to persuade the returning policeman to surrender. When his endeavours proved futile, Kelly emerged from his cover and called on the mounted police to 'bail up'. The other gang members joined him. Kennedy put his hand on his revolver. Kelly fired at him but missed. Kennedy then rolled off his horse, placing it between himself and Kelly. Scanlon, in attempting to dismount, fell to his knees and was fired on by all of the gang members. He was wounded under the right arm and fell on his side (see Figs 21.1, 21.2).

McIntyre caught Kennedy's horse and, as the gang members ran past him, he mounted it and escaped. His last observation of Kennedy and Scanlon was of both men on the ground. He heard shots being fired but did not know whether they were aimed at him.

McIntyre also gave evidence of conversations he had with Ned Kelly at the camp, before the return of Kennedy and Scanlon. They are of some significance in the ongoing debate as to whether Kelly, in shooting Lonigan, was acting in self-defence.

Fig. 21.1: Where Sergeant Kennedy was murdered. A re-enactment showing the actual tree where Sergeant Kennedy's body was found. Photo by Frederick Burman, *c.* 1880. Victoria Police Museum.

Fig. 21.2: Stringybark Creek ambush site. A re-enactment. Victoria Police Museum.

Kelly told McIntyre that he believed the police had come into the bush to shoot him. But he also said that the gang did not want the police lives, only their firearms (*Argus*, 29 October 1880).

Kelly also drew attention to the fact that the police were heavily armed and had abundant ammunition.

Earlier, at the Beechworth committal proceedings (the preliminary hearing for the trial), held in August 1880, McIntyre had deposed to asking Kelly if he was to be shot. Kelly had replied, 'No, what would I shoot you for? I could have shot you half an hour ago when you were sitting on that log if I wanted to.' What the gang wanted, according to Kelly, was the police horses and firearms (*Ovens and Murray Advertiser*, 7, 10 August 1880).

McIntyre's varying accounts

According to McIntyre's evidence at the committal proceedings, Kelly had said of Lonigan, 'Dear oh dear what a pity that man tried to get away'. His brother Dan had responded 'He was a plucky fellow, did you see how he caught at his revolver?' (*Ovens and Murray Advertiser*, 7 August 1880).

This assertion is quite contrary to McIntyre's account at the trial which had Lonigan immediately gunned down where he stood (*Argus*, 29 October 1880). On the other hand, it suggests that Lonigan was attempting to escape rather than being intent on shooting Ned Kelly.

Moreover, in his initial statement made on 27 October 1878, McIntyre had recorded: 'Constable Lonigan made a motion to draw his revolver which he was carrying. Immediately he did so, he was shot by Edward Kelly and I believe died immediately' (VPRS 1878).

This statement is, of course, in stark contrast to his trial testimony. It is also ambiguous as to whether the revolver was actually drawn before the shooting.

At the coronial inquiry into the deaths of Scanlon and Lonigan, held at Mansfield on the following day, 28 October, McIntyre deposed to yet another version of events. 'Constable Lonigan endeavoured to get behind a tree, three or four yards off – before he could do so he was shot…' (McIntyre 1878).

Kelly's subsequent claims

The time between the police shootings and the murder trial was marked by armed robberies at Younghusband's station at Faithful's Creek, at banks in Euroa and Jerilderie and the Glenrowan siege. In each instance conversations with Ned Kelly about the Stringybark Creek shootings allegedly occurred. The raid on Younghusband's station was some six weeks after the shooting and George Stephens, a groom who was held captive, stated that Ned Kelly had told him:

> Lonigan made for the log and tried to draw his revolver as he went along. He lay down behind the log and rested his revolver on top of the log and covered Dan (Kelly). I then took my rifle off (pointing at) McIntyre and fired it at Lonigan, grazing his temple. Lonigan then disappeared below the log, but gradually rose again, and, as he did so, I fired again and shot him through the head' (*Argus*, 10 August 1880).

Evidence of this conversation was corroborated by a William Fitzgerald (*Argus*, 10 August 1880).

Henry Dudley, a public servant also caught up on this raid, deposed to Kelly having with him Sergeant Kennedy's gold watch and remarking: 'It belonged to poor Kennedy. What would be best for me, to shoot the police or for the police to shoot me and carry my mangled body into Mansfield.' His evidence was corroborated by a fellow employee, Robert McDougall (*Argus*, 11 August 1880).

Another of those imprisoned at Faithful's Creek was James Gloucestor (a hawker) who gave evidence of Ned Kelly claiming to have done all the shootings. (One suspects this claim – and some other statements to people he had captured from time to time – was boasting designed to impress or cower his hostages.)

At the Beechworth committal proceedings this witness quoted Kelly as saying, 'McIntyre surrendered but Lonigan ran to a log where he was attempting to fire when I fired and hit Lonigan in the head, killing him. It was a pity he did not surrender. I did not wish to kill Lonigan, only to take their arms'(*Argus*, 10 August 1880).

At the trial the witness gave similar evidence save that, in the only reference to this conversation (in the *Argus* newspaper), Kelly is reported to have said 'Lonigan ran to the log and was trying to screen himself behind it when I fired at him' (*Argus*, 29 October 1880).

The difference in accounts may, of course, merely reflect the accuracy of the media coverage of the evidence.

On 10 December 1878, Ned Kelly told the manager of the National Bank at Euroa, Robert Scott, 'Oh I shot Lonigan.' (*Argus*, 30 October 1880).

In February 1879, when the Jerilderie Police Station had been 'stuck up' by the Kelly gang, Constable Henry Richards reported that Ned Kelly had told him, 'He had not gone out to shoot Kennedy, Scanlon and Lonigan, but was determined to get their arms. The reason he shot them was that they were persecuting him' (*Argus*, 30 October 1880).

A clerk in the Bank of New South Wales at Jerilderie, Edward Living, was given a document which Ned Kelly wished to be published. It later became known as the Jerilderie Letter, and it contained his version of the shooting of Lonigan. Namely: 'Lonigan ran some six or seven yards to a battery of logs … he had just put his head up to take aim when I shot him that instant or he would have shot me' (Kelly 1879).

The Crown witnesses who spoke with Kelly after the Glenrowan siege deposed to Kelly reiterating his by now common mantra that he had shot the police in self-defence and, if he had not shot them, they would have shot him.

Forensic pathology evidence

From a forensic medical perspective, none of the witness accounts of the shooting of Constable Lonigan are totally accurate. Dr Samuel Reynolds performed the post mortem examination of the deceased. He observed four injuries which he described as 'bullet wounds' – a superficial graze to the right temple, a wound through the right eye (which would have caused death within a few seconds), a wound where a bullet had traversed the left arm leaving a hole in it, and a wound to the left thigh where the bullet had travelled under the skin almost to the inner aspect of the thigh. Dr Reynolds told the court that the thigh wound was caused by 'an ordinary revolver bullet' (*Argus*, 30 October 1880).

On any analysis of this evidence, at least three shots were involved in Lonigan's injuries.

Advocates for Kelly's cause proffer the view that Lonigan may have accidentally shot himself in the thigh while grabbing his revolver (or immediately thereafter), before being shot by Kelly with a rifle.

The absence of any cogent, reliable ballistic evidence renders this hypothesis speculative but without doubt this material could have been utilised by Ned Kelly's defence counsel to challenge the reliability of McIntyre's account of the shooting.

The conduct of the trial

Kelly was represented at his trial by Henry Bindon, a junior barrister who had been at the bar of the Victorian colony for only 10 months and who had never appeared in the Supreme Court. His reported performance as defence counsel reflected his inexperience as a trial advocate as he underwent a baptism of fire in the crucible of the criminal trial.

Bindon made no attempt to impugn the reliability of McIntyre's evidence by reference to his prior inconsistent statements or the autopsy findings. Nor did he explore McIntyre's evidence at the committal proceedings of conversations he had with Kelly. Consequently, when McIntyre left the witness box, Kelly stood before the jury as a cold-blooded killer.

Bindon's cross-examination of the prosecution witnesses was desultory and generally ineffective. While some piecemeal references to self-defence were contributed by the Crown witnesses, the jury had no explicit or coherent account of the events and circumstances said to give rise to it.

Bindon's sole forensic achievement in the trial was successfully objecting to the tendering in evidence by the Crown of the Jerilderie Letter. In retrospect, however, this may be seen as a hollow victory. Nothing in the document was likely to increase Kelly's notorious reputation in the community. On the other hand, it set out Kelly's unhappy relationship with the police and, importantly from Kelly's perspective, it contained an assertion of acting in self-defence.

The law decreed that if the killing was of a policeman by a person seeking to resist or escape lawful arrest, the killing was deemed to be murder, unless the perpetrator could demonstrate that the policeman was acting unlawfully.

In this case it would require demonstration that the purpose of the police expedition, far from being to take the Kelly brothers into custody, pursuant to legitimately issued arrest warrants, was to summarily execute them. Unless Kelly discharged the onus which, at the time of this trial, fell upon the accused, the defence of self-defence would founder.

In light of the prosecution evidence, Kelly's task of demonstrating to the jury that he was acting in self-defence was a mammoth one.

In 1880 the law prohibited accused persons from giving evidence on oath. But Kelly was permitted to make an unsworn statement which the jury was required to take into consideration. It was absolutely essential for Kelly to provide the jury with a detailed exposition of the factors which led to the deliberate confrontation of the police party and of the immediate events which led to the shooting of Lonigan. However, his counsel called no evidence. Kelly's lips remained sealed.

The gist of Bindon's final address to the jury was that the single witness to the events at Stringybark Creek, McIntyre, was prejudiced and unreliable, and that statements of Kelly to various prosecution witnesses should be seen as designed to bluff the people in his charge [albeit some of the statements were favourable to Kelly].

He also argued that the prisoner was not a bloodthirsty assassin and that both before and after the episode at Stringybark Creek he had had many opportunities to kill police, if that was his desire. The barrister's only possible reference to self-defence in his jury address was oblique. He pointed out, 'The police had appeared on the scene, not in uniform but in plain clothes, and armed to the teeth. An unfortunate fracas occurred which resulted in the shooting of Lonigan' (*Argus*, 30 October 1880).

Sir Redmond Barry's charge to the jury, while typical of those of the 1880s, was rudimentary by today's standards (see Figs 21.3, 21.4). There was no recourse to the sophisticated legal formulas which are mandatory in the 21st century. The requirement that the Crown must prove all the elements of the offence of murder beyond reasonable doubt and negate any claim of self-defence was not definitively enunciated until the 20th century (*Woolmington v. Director of Public Prosecutions* [1935] AC 462).

From today's perspective ,Justice Barry failed to link the evidence to the relevant principles of law. The judge very briefly adverted to self-defence, when he told the jury 'as the prisoner has raised the defence of self-defence it is for him to point to circumstances arising out of the evidence which show the killing was justified' *(Argus*, 30 October 1880).

Two legal issues have attracted the attention of modern commentators.

Fig. 21.3: Sketches made during the trial of Ned Kelly, at Beechworth. *Illustrated Australian News*, 28 August 1880. State Library of Victoria.

Fig. 21.4: Judge Redmond Barry, who sentenced Ned Kelly to death. State Library of Victoria.

First, the trial judge permitted evidence of the shooting of Scanlon and Kennedy to be placed before the jury, rejecting the submission of Bindon that the shooting of Lonigan was a self-contained episode. Justice Barry told the jury that he had 'admitted the evidence of what occurred after the shooting of Lonigan because you may, if you wish, infer from it what was the prisoner's motive [probably meaning state of mind] in shooting Lonigan, or whether the shooting of Lonigan was accidental or done in self-defence' *(Argus*, 30 October 1880).

Bindon had earlier sought to have the legal correctness of Barry's ruling referred to the Full Court (three Supreme Court judges) for consideration. A decision that the judge had erroneously admitted this evidence could have resulted in a retrial.

It is likely that Barry's ruling would have been upheld on review, but in any event he refused the barrister's application. This refusal was not reviewable and in 1880 there was no right of appeal in criminal cases.

The second issue which has agitated the minds of commentators is whether self-defence was adequately placed before the jury by Justice Barry. Arguably it was not, although some sympathy may be felt for the judge given the tentative and unformed nature of the arguments put forward by defence counsel. That the trial judge failed to direct the jury appropriately on self-defence is certainly the view of the former Chief Justice of Victoria, John H. Phillips. In his book *The Trial of Ned Kelly*, he opined:

> ... the real issue for the jury in Kelly's trial was, in factual terms, the nature of the police expedition. What were the police really about? If they were bent on effecting the lawful arrest of Kelly or his brother, then Kelly's killing of Thomas Lonigan was murder; but if Kelly could show that their real purpose was to shoot him down and that in those circumstances he inflicted no greater injury on Lonigan than he in good faith and on reasonable grounds believed to be necessary in order to defend himself, then the defence of self-defence had been made out and he was entitled to be acquitted (Philips 1987).

But even if the law had been clearly stated, the meagre and fragmented nature of the evidence actually presented to the jury on this issue would have made an acquittal extremely unlikely.

Whatever the verdict in the case of Lonigan, any defence to the killing of Michael Kennedy is problematic, if the evidence of the hawker James Gloucestor of the admissions made to him by Kelly was accepted by a jury. After describing a gunfight with Kennedy, Kelly had told Gloucestor:

> He was sorry he had fired that last shot, as he thought since Kennedy was going to surrender, and not fire. He said that he had afterwards a long conversation with Kennedy, and seeing from his wounds that he could not live, he shot him. He said that the party were going to leave the ground, as he did not wish Kennedy to be torn by wild beasts while he was dying, he shot him. He added that it was no murder to shoot one's enemies, and the police were his natural enemies (*Argus*, 30 October 1880).

A claimed mercy killing would be no defence to a charge of murder.

Aftermath

In sentencing Ned Kelly to death, Sir Redmond Barry articulated the establishment view of the consequences of his conduct:

> In new communities, where the bonds of society are not so well linked together as in older countries, there is unfortunately a class which disregards the evil consequences of crime. Foolish, inconsiderate, ill-conducted, unprincipled youths unfortunately abound, and unless they are made to consider the consequences of crime they are led to imitate notorious felons, whom they regard as self-made heroes (*Argus*, 30 October 1880).

Another view may be gleaned from printed petitions circulated by Kelly supporters praying that his life be spared. In Melbourne the petitions attracted almost 35 000 signatures out of a population of 280 000 (albeit some duplications and irregularities in the petitions could be demonstrated).

In the end, only two things are certain. The life of Ned Kelly will continue to fascinate social historians and lawyers and, whatever the verdict of history, the phrase 'game as Ned Kelly' will endure as part of the Australian idiom.

Postscript

In May 2000, a modern re-enactment of the trial of Edward Kelly was held on two nights in the Victorian Supreme Court. Senior Counsel for the Crown was Julian Burnside QC, and for Kelly, Michael Rozenes QC (later Chief Judge of the County Court of Victoria). For the purposes of this presentation Ned Kelly made an unsworn statement.

I was the trial judge. I delivered a charge on each night to an audience of 300 jurors. My distinct impression was that, on the first evening, a majority would have acquitted Kelly, but on the second, that sentiment favoured conviction.

So where does that leave us? You be the judge!

References

Age, 29, 30 October 1880.
Argus, 10 August 1880.
Argus, 29, 30 October 1880.
Kelly N (1879) *The Jerilderie Letter*. Republished 2001. Text Publishing, Melbourne.
Macfarlane I (2012) *The Kelly Gang Unmasked*. Oxford University Press, Melbourne.
McIntyre, http://joeonline2.tripod.com/writings_mcintyre.htm
Ovens and Murray Advertiser, 7, 10 August 1880.
Philips JH (1987) *The Trial of Ned Kelly*. Law Book Co., Sydney.
VPRS 4966 Consignment P0 Unit 1 Item 1 Document: Report of Constable McIntyre of the murders committed at Stringy Bark Creek.
Waller L (1968) *Regina v. Edward Kelly: Ned Kelly – Man and Myth*. Cassell, Australia.
Woolmington v. Director of Public Prosecutions [1935] AC 462.

Chapter 22
Analysing the handwriting

Tahnee N. Dewhurst

Forensic handwriting analysis was conducted of several letters, including the Jerilderie letter, believed to be written by Joe Byrne or Ned Kelly, to see if the authorship of each letter could be determined scientifically.

The purpose of conducting a forensic examination and comparison of handwriting is, quite simply, to express an opinion, where possible, as to whether or not the author of one body of writing can be attributed as author of another body of writing.

Several samples of writing, using known letters from Ned Kelly and Joe Byrne, were compared to determine whether an opinion could be offered on the similarity of authorship between them and the Jerilderie letter, the Cameron letter and the Parkes letter. The five samples of handwriting analysed were:

1 a letter from Ned Kelly to Police Sergeant Babington, signed and dated 28 July 1870;
2 a letter from Joe Byrne to Aaron Sherritt, dated 26 June 1879;
3 the Cameron/Euroa letter, believed to have been dictated by Ned Kelly and written by Joe Byrne. It was sent to Donald Cameron, member of the Victorian Legislative Assembly, in December 1878;
4 the Jerilderie letter, believed to have been dictated by Ned Kelly and written by Joe Byrne, in February 1879;
5 the Parkes letter, believed to have been written by Ned Kelly, sent to the Premier of New South Wales, Sir Henry Parkes, in March 1879.

The process of forensic examination of handwriting

Upon receipt of handwritten material for analysis, examiners initially conduct a cursory or preliminary examination of the documents to ensure that the material submitted is of sufficient quality, quantity and comparative format to warrant a thorough forensic examination; if not, they reject the material as unsuitable for examination purposes. In laymen's terms this means that (a) the material is of clear and credible nature, i.e. not poor-quality photocopies, (b) enough handwriting is provided that the examiner can establish an adequate handwriting profile and the range of natural variation within a body of handwritten material and (c) the material provided for comparison purposes is similar in nature to that of the unknown or disputed writings. For instance, upper-case printed handwriting should be compared with upper-case printed handwriting or, as in this case, cursive script compared with cursive script.

Typically, handwriting is examined by looking for feature similarities and/or dissimilarities between two comparable bodies of writing. If large numbers of similarities are observed and there are no significant dissimilarities, this may support the proposition that the two separate bodies of handwriting share common authorship (were written by the same person). Conversely, if there are repeatable and inexplicable differences, this may support the proposition that the two bodies of writing do not share common authorship (were written by two different people).

This is a rather simplistic explanation of the process. There are other factors to consider, such as chance-match writing, forgery, disguised writing and any internal or external changes that might affect handwriting.

The process is predominantly based on visual observation, often supplemented by microscopic magnification. Magnification enables us to examine finer details of the handwriting, such as pen direction, letter to letter connections and more complex handwritten formations.

The process relies on the analytical and perceptual interpretation of the individual examiner. Thus, examiners should be aware of any conscious and/or unconscious bias that could affect the examination process and opinion as to authorship, and must make every effort to ignore inappropriate context information that may affect their approach to the case.

Other factors that may affect an examiner's ability to form an opinion on the authorship of two bodies of writing include:

- non-original documentation, which prevents the observation and examination of the finer elements of feature construction within the handwriting;
- unfamiliarity with 19th-century cursive script;
- limited material, particularly in the case of the Joe Byrne and Ned Kelly specimen material;
- poor-quality material, particularly in the case of the Jerilderie letter, which was of such poor quality that most pages had to be excluded, limiting the amount of handwritten material available for examination and comparison.

Our examination of the handwriting provided was limited to the observation and documentation of similarities and/or dissimilarities, due to the challenges imposed by the limited quantity and the lack of good-quality, comparable material.

Examination outcomes

The Joe Byrne specimen handwriting compared with the handwriting of the Jerilderie letter

These two documents were examined and compared; the result was that similarities and differences were observed between the Joe Byrne specimen handwriting (see Fig. 22.1) and the handwriting of the Jerilderie letter (see Fig. 22.2).

Similarities were observed between individual character formations and small repetitive words, but some differences were also noted and documented. As there were both similar and dissimilar features within the two bodies of writing, and there were limitations associated

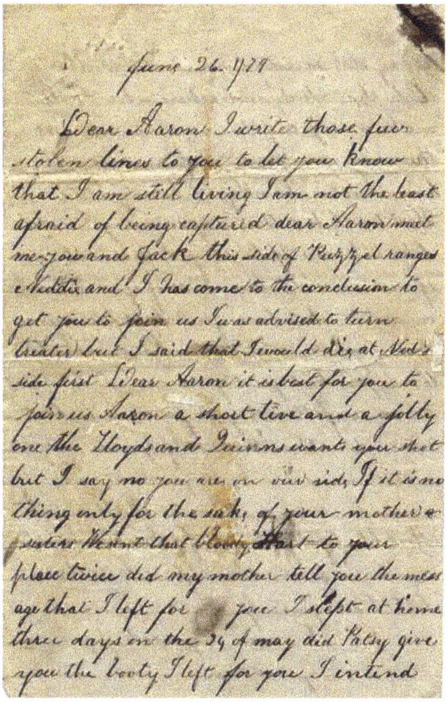

June 26. 1878

Dear Aaron I write those few
stolen lines to you to let you know
that I am still living I am not the least
afraid of being captured dear Aaron meet
me you and Jack this side of Puzzel ranges
Neddie and I has come to the conclusion to
get you to join us I was advised to turn
traiter but I said that I would die at Ned's
side first Dear Aaron it is best for you to
join us Aaron a short live and a jolly
one the Lloyds and Quinns wants you shot
but I say no you are on our side If it is no
thing only for the sake of your mother &
sisters Wasn't that bloody Wait to your
place twice did my mother tell you the mess
age that I left for you I slept at home
three days on the 24 of may did Patsy give
you the booty I left for you I intend

Fig. 22.1: A sample of the letter from Joe Byrne to Aaron Sherritt, asking him to join the Kelly gang. Public Record Office of Victoria.

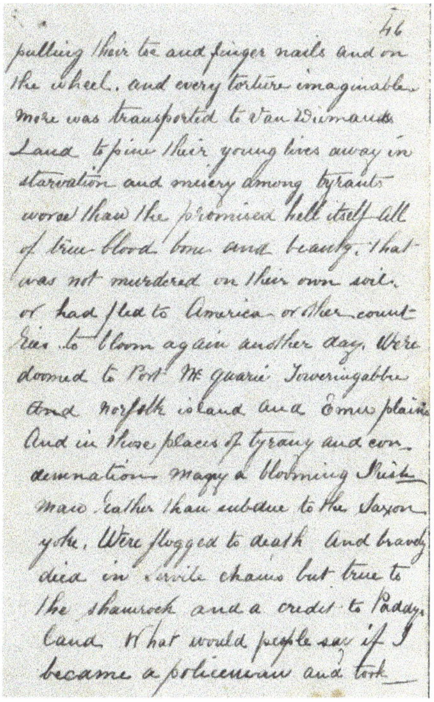

46
pulling their toe and finger nails and on
the wheel. and every torture imaginable
more was transported to Van Diemand's
Land to pine their young lives away in
starvation and misery among tyrants
worse than the promised hell itself all
of true blood bone and beauty. that
was not murdered on their own soil.
or had fled to America or other countries to bloom again another day. Were
doomed to Port McQuarie Toweringabbie
And norfolk island and Emu plains
And in those places of tyrany and con-
demnation many a blooming Irish-
man rather than subdue to the Saxon
yoke. Were flogged to death and bravely
died in servile chains but true to
the shamrock and a credit to Paddys
land What would people say if I
became a policeman and took

Fig. 22.2: A sample from the Jerilderie letter. State Library of Victoria.

Fact or fiction: There are no surviving copies of the original Jerilderie letter

The original Jerilderie letter was actually believed to have been drafted in the bush some time before the gang entered the New South Wales town of Jerilderie to rob its bank in February 1879. It was then believed to have been copied out in a neater style onto 56 pages of note paper by Joe Byrne, rather than dictated, so the original may well have disappeared.

Two contemporary copies of Joe Byrne's letter were made, one by publican John Hanlon and one by a government clerk. The Byrne letter and both handwritten copies have survived.

An earlier letter, known as the Cameron letter, along similar lines to the Jerilderie letter, was also penned by Joe Byrne during a robbery at Euroa, in December 1878. The letter was written in response to statements made by Victorian parliamentarian Donald Cameron, who had criticised the police for not apprehending the gang, the day before they were outlawed.

That Cameron letter was signed Ned Kelly and a copy was sent to both Donald Cameron and the Victorian Police Superintendent, John Sadleir. The letter largely sought to explain and justify the gang's killing of the three policemen at Stringybark Creek on 26 October of that year.

Ned Kelly hoped to have the letter transcribed at Jerilderie printed by the town's newspaper editor and printer, Samuel Gill. But he was not to be found so Kelly gave the letter to the accountant of the bank they were robbing, Edwin Living, and told him that must give it to Gill to be printed.

After the outlaws had left the town, Living set off on horseback with the letter to the nearby town of Deniliquin, to catch a train to Melbourne and deliver it to the officers of the Bank of New South Wales there. Shortly before reaching the town he stopped at the hotel of John Hanlon and allowed him to both read the letter and to make a copy of it.

After it was received by the bank officials and then reported to the police, they advised against any publication of the letter, though a summary of its content was printed. It was described by Mount Gambier's *Border Watch* of 19 February 1879:

> *It is a wandering narrative, full of insinuations and statements against the police, and of the type familiar to all who have had experience of the tales which men of the criminal stamp are accustomed to tell, it being as impossible to prevent these men from lying as it is from stealing.*
>
> And:
>
> *An account is given of the terrible tragedy at Mansfield, but it is obviously a string of falsehoods, and it would be improper for any journal to publish it.*

In July 1880 a government clerk made another copy of the letter in preparing the prosecution case against Kelly at his trial. The original letter was then returned to Edwin Living. A full copy of the letter was not made public until 1930 when it was published as a part of a series on the Kelly gang appearing in Adelaide's *Register News-Pictorial*, under the title 'The Kellys are Out', by J.M.S Davies.

Craig Cormick

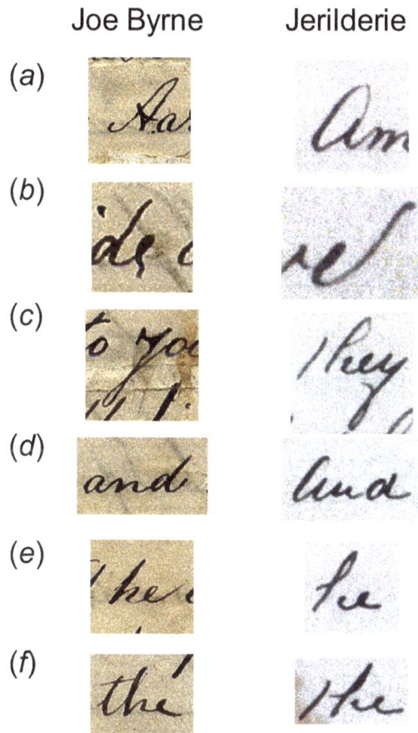

Fig. 22.3: The Joe Byrne specimen handwriting compared with that of the handwriting on the Jerilderie letter. (*a*) Upper-case 'A'; (*b*) lower-case 'e' ; (*c*) lower-case 'y'; and small repetitive words (*d*) 'and', (*e*) 'he' and (*f*) 'the'.

with the poor quality and small amount of material, we were unable to express an opinion as to authorship.

Differences observed include but are not exclusive to those listed below.

'A': Upper-case 'A' formation. An enlargement of the lower-case 'a' formation was commonly used to represent the upper-case 'A' in the Jerilderie letter but a different formation was used in the Joe Byrne handwritten material (see Fig. 22.3*a*). While not commonly employed in the material examined, the alternate formation type of the upper-case 'A' formation was also observed within the Jerilderie letter on page 48, line 13.

'e': Lower-case 'e' formation. Differences were observed between the lower-case 'e' formations at the completion of words. In the Jerilderie letter the handwritten trace travelled upwards at the end. In the Joe Byrne specimen material the lower-case 'e' formation traversed downwards at the end (see Fig. 22.3*b*).

'y': Lower-case 'y' formation. Differences were observed between the construction and formation of the lower-case 'y' formation. In the Jerilderie letter the lower-case 'y' had an anti-clockwise looping formation at the base; conversely a clockwise looping formation was observed at the base of the character in the Joe Byrne material.

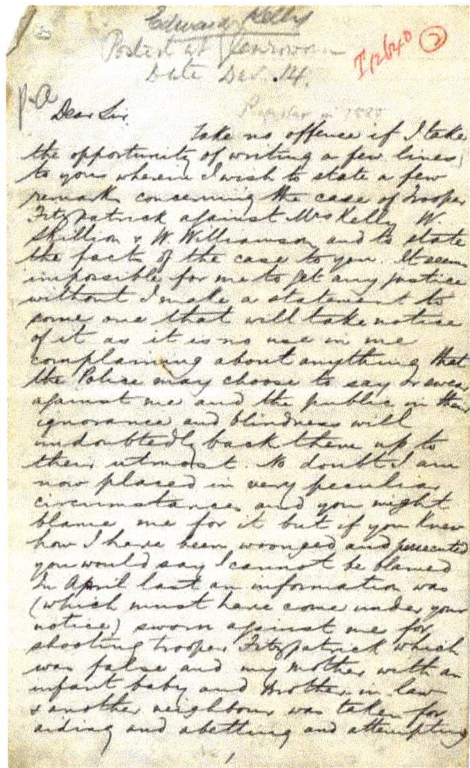

Dear Sir,
Take no offence if I take
the opportunity of writing a few lines
to you wherein I wish to state a few
remarks concerning the case of trooper
Fitzpatrick against Mrs Kelly W.
Skillion & W. Williamson and to state
the facts of the case to you. It seems
impossible for me to get any justice
without I make a statement to
some one that will take notice
of it as it is no use in me
complaining about anything that
the Police may choose to say or swear
against me and the public in their
ignorance and blindness will
undoubtedly back them up to
their utmost. No doubt I am
now placed in very peculiar
circumstances and you might
blame me for it but if you knew
how I have been wronged and persecuted
you would say I cannot be blamed
In April last an information was
(which must have come under your
notice) sworn against me for
shooting trooper Fitzpatrick which
was false and my mother with an
infant baby and brother in law
& another neighbour was taken for
aiding and abetting and attempting

Fig. 22.4: A sample from the Cameron/Euroa letter. Public Record Office of Victoria.

Additionally, the body of lower-case 'y' in the Joe Byrne specimens was shallower than that observed in the Jerilderie letter (see Fig. 23.3*c*).

Anomalies were also observed with respect to letter relationships within wording, e.g. 't' to 'h' combinations, and there were construction and formation variations within small repetitive words such as 'and', 'he' and 'the' (see Fig. 22.3*d*, *e*, *f*).

This examination and comparison was limited by the poor-quality, non-original documentation (most of the Jerilderie letter is of such poor quality as to warrant exclusion), examiner unfamiliarity with 19th-century handwriting, limited material (Joe Byrne specimen letter is limited in quantity) and unfamiliarity with anomalies of 19th-century writing implements.

Opinion: Similarities and dissimilarities were observed. Inconclusive as to common authorship.

The Joe Byrne specimen handwriting compared with the unknown handwriting of the Cameron/Euroa letter

These two bodies of handwriting were examined and compared with the result that, in my opinion, there is an overall pictorial dissimilarity.

The handwritten entries in the Cameron/Euroa letter (see Fig. 22.4) appear to be more fluently executed than the handwritten entries in the Joe Byrne letter.

Fig. 22.5: The Joe Byrne specimen handwriting compared with that of the handwriting on the Cameron/Euroa letter. (*a*) Upper-case 'A'; (*b*) lower-case 'g'; and small repetitive words (*c*) 'and', and (*d*) 'the'.

Differences observed include but are not exclusive to those listed below.

'A': upper-case 'A' formation. An enlargement of the lower-case 'a' formation is utilised in the Cameron/Euroa letter but a different formation is used in the Joe Byrne handwritten material (see Fig. 22.5*a*).

'g': Lower-case 'g' formation. The lower-case 'g' formation predominantly executed within the Cameron/Euroa letter lacks a curved tail to the base, compared with the lower-case 'g' formation in the Joe Byrne material, which includes a curved tail (see Fig. 22.5*b*).

Small repetitive words such as 'the' and 'and' (see Fig. 22.5*c, d*).

Opinion: Dissimilarities were observed. Inconclusive as to common authorship.

The Ned Kelly specimen handwriting compared with the handwriting of the Jerilderie letter

The handwritten entries in Ned Kelly's letter to Sergeant Babington (see Fig. 22.6) and the Jerilderie letter were examined and compared. There were similarities and dissimilarities between the formation and construction of the handwriting in the two documents.

Two of the similarities are described below.

'h': lower-case 'h' formation. A two-stroke lower-case 'h' formation was observed in both Ned Kelly's letter to Sergeant Babington and the Jerilderie letter (see Fig. 22.7*a*).

'and': a pen lift between the 'n' and the 'd' in the word was observed in both bodies of writing (see Fig. 22.7*b*).

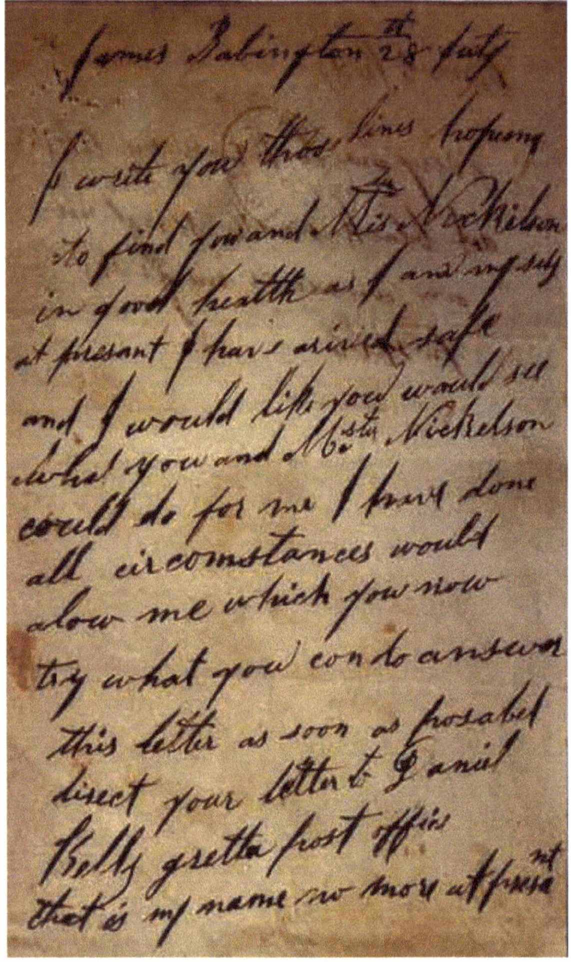

James Babington 28th July
I write you thos lines hopeing
to find you and Mis Nickelson
in good health as I and myself
at present I have arived safe
and I would like you would see
what you and Mstr Nickelson
could do for me I have done
all circomstances would
alow me which you now
try what you can to answer
this letter as soon as posable
direct your letter to Daniel
Kelly gretta post office
that is my name no more at present

Fig. 22.6: A sample from Ned Kelly's letter to Sergeant Babington. Public Record Office of Victoria.

Fig. 22.7: The Ned Kelly specimen handwriting from the Babington letter compared with that of the handwriting on the Jerilderie letter. (*a*) Lower-case 'h' formation (two-stroke formation, with a pen lift between the vertical stroke and the body of the formation); (*b*) pen lift between the letters 'n' and 'd' in the word 'and' (note length of vertical staff in the lower-case 'd' formation and the angular discrepancy caused by the height relationships between letters.

It should be noted, however, that the examination was limited by the small amount of handwriting available for examination and comparison within Ned Kelly's very short letter to Sergeant Babington, and the nine years between the writing of the two letters. Ned Kelly's letter to Sergeant Babington was purported to be written in 1870 and the Jerilderie letter is documented as being written in 1879.

Opinion: Similarities and dissimilarities observed. Inconclusive as to common authorship.

The Ned Kelly specimen handwriting compared with the handwriting of the Cameron/ Euroa letter

The handwritten entries within Ned Kelly's letter to Sergeant Babington and the Cameron/ Euroa letter were examined. Dissimilarities were observed between the formation and construction of the handwritten formations within the two documents.

However, again it should be stated that there were limited amounts of handwriting to compare and an eight-year separation in the texts.

Opinion: Dissimilarities observed. Inconclusive as to common authorship.

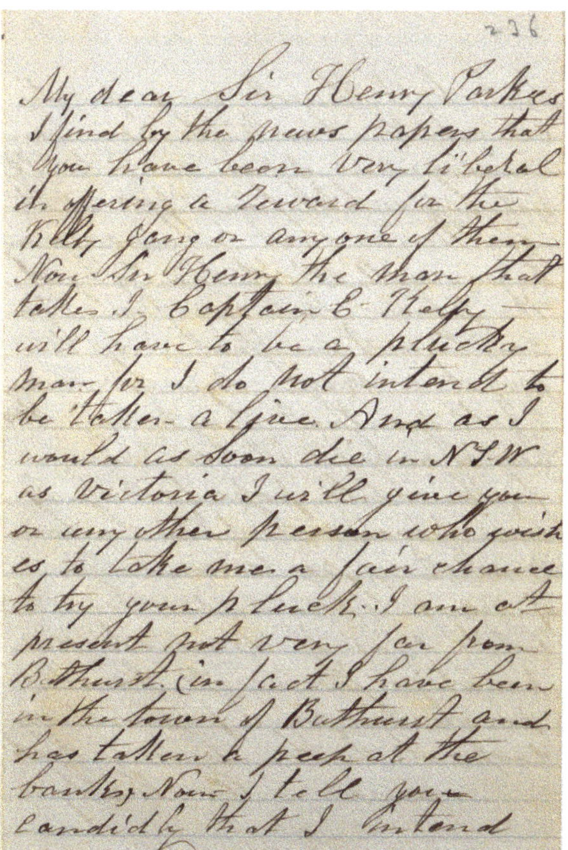

236

My dear Sir Henry Parkes
I find by the news papers that
you have been very liberal
in offering a reward for the
Kelly Gang or anyone of them
Now Sir Henry the man that
takes I Captain E Kelly
will have to be a plucky
man for I do not intend to
be taken alive. And as I
would as soon die in NSW
as Victoria I will give you
or any other person who wish
es to take me a fair chance
to try your pluck. I am at
present not very far from
Bathurst in fact I have been
in the town of Bathurst and
has taken a peek at the
banks. Now I tell you
candidly that I intend

Fig. 22.8: A sample from the Parkes letter. State Library of New South Wales.

Fact or fiction: Ned Kelly was illiterate

There are enough surviving examples of Ned Kelly's handwriting to know that he could write. The reasons there is some belief that he may have been illiterate are that the Jerilderie letter was penned by Joe Byrne, and that Ned signed his gaol letters with an X.

The reason Joe Byrne wrote the Jerilderie letter was probably his neater penmanship, and while in gaol Ned Kelly was unable to write due to a wound he received in his thumb at the shoot-out at Glenrowan five months earlier, so he had to dictate his letters.

His final letter was dictated in his gaol cell on 10 November, the day before he was executed. It was witnessed by William Buck, a warder, and was addressed to the Governor of Victoria, George Augustus Constantine Phipps, addressing him as His Excellency the Marquis of Normanby, and signed with an X. It ends with:

> *I know now it is useless trespassing on your valuable time because of the expense the Government has been put to which is not my fault. They will only be satisfied with my life, though I have been found guilty and condemned to death on a charge which, of all men in the world, I should be the last one to be guilty of.*
>
> *There is one wish, in conclusion, I would like you to grant me, that is the release of my mother before my execution as detaining her in prison could not make any difference to the Government now, for the day will come when all men will be judged by their mercy and deeds; and also if you would grant permission for my friends to have my body that they might bury it in consecrated ground.*
>
> *Edward Kelly*

Another letter, whose authenticity has been disputed by some but verified by others, was written to the New South Wales Premier, Sir Henry Parkes, in March 1879, in Ned Kelly's own hand. It was written, it seems, to unsettle the Premier.

> *To Sir Henry Parkes*
> *Premier NS.W*
>
> *My dear Sir Henry Parkes*
>
> *I find by the newspapers that you have been very liberal in offering a reward for the Kelly gang or any one of them. Now Sir Henry the man that takes I Captain E. Kelly will have to be a plucky man for I do not intend to be taken alive. And as I would as soon die in NSW as Victoria I will give you or any other person who wishes to take me a fair chance to try your pluck. I am at present not very far from Bathurst (in fact I have been in the town of Bathurst and has taken a peep at the bank). Now I tell you candidly that I intend to rob Bathurst and particularly the bank. So now you are warned of course I will not say what time I and the gentlemen that follows in my train will visit the City of the plains. But one thing you can count on that I will pay it a visit. Now Sir Henry I tell you that highway robbery is only in its infancy for the white population is been driven out of the labour market by an inundation of mongolians and when the white man is driven to desperation there will be desperate times. I present my respects to the Sydney police*
>
> *yours E. Kelly*

Craig Cormick

The Ned Kelly specimen handwriting compared with the handwriting of the Parkes letter

The handwritten entries in Ned Kelly's letter to Sergeant Babington and the Parkes letter (see Fig. 22.8) were examined and compared. Similarities were observed between the formation and construction of the handwriting in these two documents.

The examination was again limited by the small amount of handwriting available for examination and the eight-year discrepancy between the two letters.

Opinion: Similarities observed. Inconclusive as to common authorship.

Chapter 23
Managing the news: a personal perspective

Deb Withers

Managing the information about the forensic analysis of Ned Kelly's remains was a major logistical feat as interest in the story was global and sometimes frantic. Keeping the story under control required careful planning, close collaborations and tight management of information to prevent incorrect stories getting into the public domain.

One morning in September 2011, the world woke up to find Ned Kelly in their lounge rooms, in their cars or at their breakfast tables.

Despite the notorious bushranger's demise more than 130 years previously, Ned was still making headlines and someone had to wrangle the media to try and ensure that the story was told accurately and well. That task fell to me.

For as much as forensic sciences demand close attention to detail and rigorous checking of the facts, so did the release of information about the forensic work the Victorian Institute of Forensic Medicine was conducting. And, besides, it's not every publicist who gets to list Ned Kelly as a client!

As media adviser to the Institute I had foreknowledge that we were about to admit the human skull purported to be that of Ned Kelly, before Tom Baxter brought it in, in an old, faded and battered beauty case. That was on 11 November 2009, the 129th anniversary of Ned's execution.

At first there didn't appear to be much enthusiasm around the Institute. Human remains are the Institute's day-to-day business after all, and some of the forensic experts could see no difference between this skull and that belonging to any other poor soul.

But I knew there was going to be great human interest in the story. In cold, calculating terms this was going to be a great story and, whichever way it ended, the VIFM would attract massive coverage. But the information had to be handled as delicately as the human remains had to be handled, as Ned has a history of generating myths and misinformation.

The incumbent Attorney-General, Rob Hulls, was easy to get on side. With his Irish background and love of history and literature, Rob also had a reputation as a bit of a larrikin. And that was perfect for being the initial spokesperson for Ned.

He fronted a press conference the day after the skull was delivered and came out with some great lines like 'Is this Ned's head or just another dull skull?' He proved to be pretty much a publicist's dream.

The headline of the *Age* newspaper the following day read:

> Experts to examine 'Ned Kelly' skull (*Age*, 13 November 2009).

It was on the cover of most of the dailies and the *Herald Sun* featured a cartoon of Rob Hulls asking a laboratory coated scientist with a magnifying glass examining the skull if it was 'Ned's noodle'. The skull replied 'What're you looking at you big ugly, fat-necked, wombat headed big bellied, magpie legged, sons of Irish bailiffs!'

In PR terms, if your client is depicted in a cartoon, you know you've really made it!

That was a much better result for the Institute and the reputation of good science than those that read:

> Australian farmer claims skull is Ned Kelly's (*Guardian*, 13 November 2009).

The story was running. And over the next 22 months I not only monitored and analysed and drafted material for release to the media and the public, but juggled, cajoled, manoeuvred, appeased, schmoozed and strategised to keep it under control.

There was a change of government, negotiations with a production company and broadcaster, information kept from an insatiable press, meetings with the Kelly family and a cast of academics to work with.

I learned a lot more about mitochondrial DNA, CT scanning, imaging and craniofacial photo superimposition than my poor unfortunate science teacher Mr Trotter could have ever hoped to teach me, and I was reintroduced to a piece of Australian history that I had avoided whenever possible at school. But I needed to be aware of every aspect of the story to ensure that what was being run in the media and online was both accurate and in the best interests of the Institute.

As a potential once-in-a-lifetime opportunity for a forensic institute that largely hides its light under a bushel and only rears its head when the issue is forced, I could see this would put us on the world stage. And as we are a public institute, funded by the taxpayer, we need to let the people of Victoria know what we are up to. This was a great opportunity to profile our work, as it combined science and history – subjects for which the public has a seemingly unquenchable thirst – especially since the popularity of shows like *Silent Witness* and the *CSI* franchise.

But public perception of how forensic science works, often as result of such TV shows, is often oversimplified or oversensationalised. Here we had an opportunity to explain the science to people in a bit more detail and increase public understanding of its workings. For instance, forensic analyses can be slow and painstaking and the results can sometimes be far from certain – a different world from the instant and accurate results that forensic scientists seem to be obtain on CSI TV shows.

Within the Institute as well it provided an opportunity for people with different skills to work together on a joint project, and to gain some more understanding and respect for each other's areas of expertise.

Interest in Ned Kelly is so strong we could expect to see magazine covers, page one in the papers, lead stories on radio and television, blanket online coverage and even a book and a documentary. If managed well.

Books, documentaries, magazines, chat show interviews, scientific journals and ancillary media don't just happen overnight. They all took a lot of organisation and strategic planning. Release dates needed to be coordinated and the timing of announcements had to be just

Fact or fiction: Ned Kelly was more hero rather than villain

This is perhaps one of the most divisive questions about Ned Kelly, and it is easy to either lionise or demonise him. To some he was a Robin Hood figure who was a victim of the oppression of the authorities of his time and stood up for the underdog in fighting back. To others he was a criminal and a murderer.

Many people's view of Ned Kelly is framed from their own social perspectives, and any understanding of Ned Kelly the man tends to disappear as a result of it.

He has been described repeatedly as 'one of Australia's greatest folk heroes' and a standard expression of bravery is 'as game as Ned Kelly'.

But to the descendants of the police killed by him and his gang at Stringybark Creek, and to many members of the police force since, he has been described as 'a psychopathic criminal misfit who left a trail of destruction and misery in his wake'.

But is either position an oversimplification of the man? Can he actually be both?

He was certainly a product of his background, his times, his family upbringing and the society he mixed with. His father, a former convict and ongoing cattle thief, died when Ned was 11 or 12 years old, having contracted a disease in prison. This left Ned, not yet a teenager, the oldest male of a family of seven children.

His family and close relatives had numerous run-ins with the police and Ned was first arrested at the age of 14 on a rather dubious case of assault on a Chinaman.

It is clear that Ned and his family, from the very earliest, had a vendetta against the police and it is also clear the police had something of a vendetta against the Kellys. Given the mutual animosity, an escalation of the conflict on both sides was inevitable.

It is true that Ned Kelly was very hostile and violent towards the police, but it is also true that he went on the run after being accused of attacking a policeman, Constable Fitzpatrick, who most likely invented the story, being later thrown out of the police force for drunkenness and perjury.

Ned's mother was imprisoned in his place – undoubtedly fuelling his outrage as expressed in the Jerilderie letter:

> *… and yet the ungrateful articles convicted my mother and an infant, my brother-in-law and another man who was innocent, and still annoy my brothers and sisters … This sort of cruelty and disgraceful and cowardly conduct to my brothers and sisters who had no protection, coupled with the conviction of my mother and those men certainly made my blood boil. I don't think there is a man born could have the patience to suffer it as long as I did, or ever allow his blood to get cold while such insults as these were unavenged.*

Opinions on Ned Kelly were as much divided in his day. He was roundly criticised in the press as a violent criminal and yet when he was sentenced to be hanged, 60 000 people signed a petition asking that he be spared.

Craig Cormick

right. There also needed to be an extent of drip-feeding to assuage a press that is ever-curious about Ned, without giving away all the information at once.

Deadlines had to be adhered to, confidences kept, various aspects of the story shared appropriately according to the media's audience and reach, and there was a degree of quid pro quo happening constantly. I've worked with a few blokes who are recognised by just their first names – Elton, Molly, Eddie and Bert – and while they are all still inclined to occasionally cause trouble, none of them have the divisive effect on the nation that Ned does.

This presented its own challenges. No-one at the Institute could be seen to take sides over the hero/villain argument. The team identifies with impartially delivered justice, works with Victoria Police and serves the public. So while they all obviously had their own opinions they had to be even-handed, and seen as such.

It was fascinating to watch members of the team's interest grow. Some members slowly, and then more quickly, began to take a strong interest, realising the attraction that this project would generate made it important for the profile of the Institute as well.

There were constant calls from many members of the press and the fielding of questions was often tough. When it comes to Ned, everyone has an opinion. Everyone wants a piece of him. And despite many people having some sense of proprietary ownership, he really belongs to all Australians. He's part of our folklore.

This made it difficult to strategise the best outcome for the Institute as everyone had a strong feeling on how the story should run. For instance, we'd negotiated to produce a documentary for SBS and that wasn't always easy. The crew came from Perth so many of the forensic procedures had to be timed to coincide with their shooting schedule.

Then the airing date of the documentary was put back several times, which meant our announcements had to be delayed as well. It became a tricky balancing act. *Ned's Head* finally screened on Sunday 4 September 2011.

Straight after the announcement of the identification, there was also the Kelly family to consider, respect and protect. Again, hungry for any morsel, the media plied them and tried anything to get a story. Many of the family still live around the Glenrowan area and have, over the years, become very cautious of talking to the media. Ned has had a huge impact on their lives for more than 140 years – generally speaking, they have become very private and protective as a result of that.

It's the media's job and it used to be mine, so I know how it works, with the need to get a story before your competitors being crucial. But it's not easy for those who don't live in that world or seek that attention. It can become quite intimidating.

To try and appease the public's need for information we established a special website on the Ned Kelly Project on the Institute's website (www.vifm.org), with explanations of DNA and profiles of the forensic scientists involved. There was also a page for members of Ned's family to lodge their views on what should be done with his remains.

Announcing the results

After almost two years, on 1 September 2011, the day of the press conference to announce our findings finally arrived. Even for cynical and experienced old me, it was a big day. I underestimated just how far afield Ned's appeal reached. As soon as the announcement was made that although the skull purported to be Ned was not Ned, yet we had identified the rest of Ned, the phones rang hot.

It was the quintessential good news and bad news scenario.

I would hang up from one call to find three messages on my voicemail. Ned was huge and was worldwide. The story was on the BBC, CNN, Al Jazeera and other international networks, and of course all Australian broadcasters ran with it. We were on the front page of newspapers everywhere: London, New York, Tokyo, Dublin, Paris, Rome … together, Ned and the Institute blanketed the world.

Some of the headlines were memorable, including:

> Bones of Aussie Outlaw Rise Again (*Boston Globe*)
>
> Historic Outlaw Still One Step Ahead (*Irish Times*)
>
> Let Ned RIP (*Border Mail*, 22 May 2011)
>
> Not Ned's Head but a ripper yarn all the same (*Sydney Morning Herald,* 6 September 2011)
>
> Bones of Australia's Jesse James are identified, but his skull remains a fugitive (*New York Times,* 1 September 2011)

And there was another cartoon in the *Herald Sun.*

It was quite extraordinary to think a young Irish-Australian man whose father was born in a district not far from where my mother's family came from, could still have such an impact on the world.

It took some time for the phones to finally slow down, and the stories continued right up until Ned's remains had been handed over to his family and buried in an unmarked grave at Greta in January 2013. But looking back the media coverage was generally good and by and large accurate. Both Ned and the Institute had done well.

On a personal note, Ned had bridged a gap between the past and the present, and there are some special 'Ned moments' that will always be mine. He had given me some very special opportunities and special memories, such as the chance to work more closely, and in a different capacity, with colleagues at the Institute. I will always remember, for instance, the fascinating session I spent in the mortuary with Chris Briggs, our forensic anthropologist, when he explained Ned's injuries in detail. It was important that I had this knowledge.

And he had given me cause to revisit an earlier Ned moment from my own past – my Nanna Annie Brices' stories about Ned and 'the boys' watering their horses at Nanna Grant's farm near Wangaratta.

The Ned Kelly story for the Institute has now been told – for me, the drive up 'the Hume' through Kelly country will never be quite the same again.

Chapter 24
The end of a 70-year journey?

Ian Jones

Ian Jones has spent 70 years researching Ned Kelly's life. The rare opportunity to view his skeletal remains proved a moving one, but also enabled Ian to learn what the skeleton could tell about the battle with police at Glenrowan.

Feeling awkward in a green medical gown and blue overshoes, I was grateful that my similarly garbed companion didn't look as odd as I felt. Nor did our guide, Dr Soren Blau, for this was her usual work uniform as one of the key group of Victorian Institute of Forensic Medicine's scientists who had spent more than a year sorting Ned Kelly's bones from 32 disassembled skeletons excavated from a burial ground at Pentridge Prison. It was a stunning achievement.

In a definitive moment, Dr Blau ushered us into the homicide room of the Institute with its wet, white-tiled floor and row of viewing windows. I was unaware of such details at the time; I saw only the autopsy trolley in the centre of the room, its stainless steel top masked by a pale blue board on which a collection of old, brown bones were arranged in a skeletal form that was strangely dehumanised by the lack of head and hands. Yet the first impact was of physical strength and a power that almost outweighed the significance of the moment.

It was 15 May 2012. With the approval of a committee of Kelly descendants and the VIFM we were about to confront all that remained of Ned Kelly – or all that had been identified of him. The bones, carefully placed in their original articulation, had been identified by their perfect match with the DNA of the green-gowned man at my side, my friend Leigh Olver, Ned Kelly's great-grandnephew. Together we adjusted to the strange unreality of the moment – the sanitised impersonality of the setting and the unexpected heart-clenching power of the incomplete skeleton – before we moved in to examine what we could read of the man's life implicit in this, its ultimate artefact.

For Leigh this was the culmination of a lifetime's search for contact with his remarkable ancestor. For him, the saga of Ned Kelly's life that had enthralled me since childhood was a recital of family tragedies shared with his grandmother, for whom Ned and Dan Kelly were Uncle Dan and Uncle Ned, and Mrs Kelly was Granny. For me, this seemed the end of a 70-year journey that had started in 1941 when I was 10 and read two books about Ned Kelly, in one of which he was a consummate hero and in the other the darkest of villains.

By the time I left school I had read a dozen books, prowled through Kelly-era newspaper files and was studying the 17 000 questions and answers of the Royal Commission of 1881 into the Kelly Outbreak. University years introduced me to archival research and historical discipline. My journey continued through careers in journalism, television and film, all contributing. It was Philip Adams or Barry Jones who once said that my quest was something between a career and a religion.

What could the skeleton tell us?

What were my priorities as I approached the trolley? In my 70 years of reading and searching, I had catalogued the 28 gunshot wounds that Ned had suffered in the police siege of the Glenrowan Inn – the most serious wounds in the very start of the gun battle at about 3.20 a.m. on Monday 28 June 1880. The last wounds brought him down at around 7.45 the same morning. I was about to learn what the skeleton could tell us about those pre-dawn and sunrise battles.

As a party of police charged towards the Glenrowan Inn in brilliant moonlight, Ned Kelly in full armour stood on a roadway 6–8 yards in front of the pub's verandah where the other armoured gang members also stood with their guns ready. As the police leader, Superintendent Hare, barged through a revolving gate about 30 yards from the inn, Ned fired the first shot of the battle from his Colt revolving rifle. The bullet put Hare out of action with a smashed wrist. Moments later a shot almost certainly fired by Constable Gascoigne should have put Ned out of action. Ned was facing Gascoigne, aiming his rifle, so the bullet drilled through his bent left forearm then through the upper arm. The single shot had inflicted four wounds – entry and exit wounds in both the lower and upper arm. The limb was effectively shattered.

Before he could recover, a second bullet, also probably fired by Gascoigne, struck Ned's right foot by the big toe. Gascoigne was using a Martini-Henry rifle, a formidable military weapon that fired a 0.45-calibre bullet with an effective range of 800 yards. In recent years I had examined Ned's boot in the State Library. The gaping hole in the toe was obvious but there was no exit wound at the heel nor any sign of damage. It seemed unbelievable that this projectile, only 30 yards into its trajectory, could be stopped by the flesh, muscle and bone of a human foot.

Finding the incredible truth of the injuries

The truth, as reported by VIFM scientists, was even more incredible. I knew from early press reports, only days before our viewing, that the bullet had split in two and lodged among the bones of the big toe. As I prowled around the far end of the trolley, my eye was drawn to the two dark and incongruous knobs on the right foot – misshapen fragments of lead actually fused with the bones of the big toe. Such was the result of Gascoigne's second shot.

I now turned from the first wounds to the last wounds, when Sergeant Steele fired two charges of shot into Ned's legs at close range and brought him down. Dr Blau pointed to a perfectly circular hole in the right tibia 8 mm in diameter and showed us two tiny corroded and blackened pieces of lead that had once been an exact fit for the hole and rattled around inside the bone when it was recovered.

When I read a report of this find, I commented, 'This wasn't shot – it was shrapnel.' Today, 'shrapnel' is taken to mean shell fragments or any metal flying through the air after an explosion. True shrapnel, named after its inventor Englishman Henry Shrapnel and used widely by all armies in World War I, consisted of musket balls loaded into a shell and blasted downwards on the enemy by a time-fuse as the shell was descending. In shooting circles, shot of this size is called SG ('small grape'), loaded nine pellets to a cartridge. Incredibly, even though at the time of Ned Kelly's Beechworth trial it was reported that he had 23 wounds 'in the lower extremities', that one hole in the right tibia is the only bone damage that had been found on his legs. Even if the other 16 pieces of SG had lodged in soft tissue, it seems hard to believe that they had all missed bones. Significantly, the angle of the wound showed that Sergeant Steele had been half behind Ned when he felled him. Again my eye flashed to the hideously crippled right foot.

I turned to move along the other side of the trolley to see more closely the effect of Gascoigne's first shot, but I already knew that most of the evidence had disappeared. The upper bone of the left arm showed a deep and broad gouge from just above the elbow diagonally to where the bullet had exited just below the shoulder. The surprisingly delicate lower bones of the left arm – the radius and ulna – were shortened by post mortem amputation a little below the elbow. The face of the cuts were darkly aged in sharp contrast to white scars where bone had been cut away for DNA sampling. Remnants of the bones also showed signs of infection.

The skull fragment

Above, where the skull should have lain, there was one fragment scarcely the size of my palm – a broad, blunt and down-pointing arrowhead of slightly dished bone with white-cut upper-edges marking another harvest of DNA, its dark outer sides slanting to its point blunted by a semi-circular scoop to accommodate the top of the spine.

Dr Blau pointed to brown saw marks on the top vertebrae that aligned with the old-cut edges of the skull fragment, showing that spine and skull had been connected when the fragment was cut out. These saw marks were on the third vertebrae but the first and second cervical vertebrae – the atlas and axis – were missing. Stubbornly clinging to my belief that Ned Kelly's head had been removed after hanging, it seemed credible to me that these two vertebrae had been destroyed during decapitation. Some teeth found with the skeleton could have suggested that at least part of the skull had been buried with it, but the DNA of the teeth was from another person.

Now Dr Blau lifted two vertebrae from their careful alignment and pointed to the unusual degree of wear and degeneration in what other markers confirmed as a young and unusually robust frame. VIFM scientists had seen this as being caused by Ned's 'hard riding'. I looked at this middle-aged spine on a 25-year-old man and read teenage work as a splitter and fencer, then three years' hard labour in Beechworth Gaol, Pentridge and on a hulk at Williamstown. From his honest years I saw work as 'an expert axeman' with sawmills in north-east Victoria and Gippsland. I saw regular winter shearing at Gnawarra station on the Darling. I saw him quarrying and cutting granite blocks to build a homestead beside Lake Winton. It all bore out the rare praise from one of his harshest critics, Beechworth's *Ovens and Murray Advertiser*: 'No man could work harder or better than he when he chose.'

Never harder and better than in 1877, the last of his honest years, when he built a bark-and-slab homestead for his mother. He felled the trees, cut, split and adzed hundreds of ironbark slabs, then manhandled and stacked them between the studs of the frame. One smaller slab, 1.5 m long by 45 cm wide and 7.6 cm thick, weighed 18 kg. And on top of all this, there was of course plenty of hard riding.

My eyes strayed again to the ugly deformity of the right foot and marvelled at the man who could bear that wound and the shattered left arm and go through four hours of blood loss and trekking through the bush to and from the Glenrowan Inn, weighed down by 44 kg of armour, before he came back at sunrise to attack the 34 police who had the surviving gang members surrounded in the inn. All this had been mere prologue to a half-hour gun battle in which another 17 bullets hit his armour and he suffered several more wounds before the two shrapnel bursts in the legs brought him down.

> When he started the fight he was already in a state of advanced hypovolemic shock, the last 18 wounds tipping the balance; in Ned's own words, 'I was unable to stand and the helmet I wore was choking me and the bullets completely stunned me.' (Constable W. Phillips 1880).

This was an inside-the-helmet view of near-fatal blood loss. He crashed down on his knees as a despairing cry echoed from the helmet: 'I'm done! I'm done!' (*Age*, 29 June 1880).

What he had done seemed to me impossible. Beyond endurance, beyond courage, beyond anything that could be expected of a man in his condition, and it took place in subzero temperatures after he had gone at least two nights without sleep. It was, in the truest sense of the word, superhuman.

When Leigh and I finally left the room, I glanced back at the assembled bones on the pale blue board. The first sight of them had occupied a moment of indescribable power and mental hyperventilation; this last glimpse as the door closed was woven with new revelations and ongoing mystery. I discarded my green gown and blue overshoes and with them the guise of historical clinician. I had at last confronted the ultimate artefact of Ned Kelly's life, yet I had no sense of fulfilment, of completion. I knew that my search was not over. I had encountered a reality that overtowered the legend.

My search was reinvigorated into its eighth decade.

References

Age, 29 June 1880.
Constable W. Phillips, claim to Kelly Reward Board, 20 December 1880.

Chapter 25
So who has Ned's head?

Craig Cormick and Fiona Leahy

There are several theories as to what may have become of Ned Kelly's skull – or to be precise, the rest of his skull, since a portion of it was recovered with his remains. One theory is that it was never reunited with the body that was buried at the Old Melbourne Gaol. This theory is based on the belief that the head was severed for making the death mask, rather than the mask being made by placing the head through a hole in a desk to work on.

There are certainly many unconfirmed stories that it was used as a paperweight, somewhere, by someone.

The author Jill Dimond proposed that it was taken by the phrenologist Archibald Hamilton (see p. 126), to add to his collection of skulls (Dimond 2013). She argued that Hamilton had a track record for going to any lengths to get skulls and death masks for his collection, and had been taken to court charged with stealing them on several occasions. Hamilton had in fact applied for permission to take phrenological readings of Kelly's skull. The job of making the death mask had been granted to Maximilian Kreitmayer, who allowed Hamilton to attend.

If Hamilton did end up with the skull he kept it secret and, when he died in 1884, Agnes, his wife and business partner, never revealed its whereabouts. Dimond stated that it may have 'ended up in the collection of another phrenologist, or been acquired by a museum ('We have cases of them in the basement') or simply discarded as worthless' (Dimond 2013).

Another claim for the skull has been made by Anna Hoffman, a New Zealand woman in her 70s, who had courted controversy as a witch in the 1960s and 1970s. She has said she has around 20 skulls in her possession and was on holiday in Melbourne in the 1980s when a she met a security guard who gave one to her. She said he told her it was the skull of Ned Kelly. From photos of the skull that have been published, this is not likely.

Another less exciting possibility is that it was buried in the graveyard at the Old Melbourne Gaol, but was crushed during the excavations in 1929.

Whoever does, at any point, claim to have the skull, it will be apparent from a cursory glance if the claim is to be taken seriously by examining the back of the skull, to see if a chunk has been cut out that fits the piece of skull found with Ned Kelly's remains.

During the course of its investigations the Victorian Institute of Forensic Medicine had to address many theories and beliefs put forward about Ned Kelly's (the 'Baxter') skull. The main ones are addressed below.

1 The 'Baxter skull' is Frances Knorr's, as someone altered the grave-marker from FK to EK.
The Institute has confirmed that the skull was from a male. It also compared it to the death mask of Frances Knorr – it was not a match.
2 The 'Baxter skull' is Ernest Knox, as he had the same initials as Edward Kelly and could have been buried under the EK grave-marker.
There were two similar grave-markers on the wall of the Old Melbourne Gaol: E.K. with the broad arrow and E.K with 19.3.1894. The latter is still at the Old Melbourne Gaol museum and belongs to Edward Knox. A comparison was made between the 'Baxter skull' and the death mask of Ernest Knox and it did not match (see Fig. 24.1).
3 Kelly's body was dissected after execution, his head was removed, portions of his corpse were taken and put into curiosity cabinets, and his skull and brain were souvenired and are now in private hands.
This view is based on the *Bendigo Independent* article of 1890 that stated: 'Ned Kelly's body, (writes our gossipy correspondent) was given over after the execution to the medical men, and a nice mess, I am told, they made of it. The students particularly went in heavily taking part of his body and generally examining every organ … I am told that portions of the corpse are now in nearly every 'curiosity' cabinet in Melbourne medical men's places. The skull was taken possession of by one gentleman, and it is probable that he may hereafter enlighten us upon the peculiarities of the great criminal's brain.'
However, the police documents discovered by Helen Harris undermine the veracity of this report (see Chapter 2). The saw marks on Kelly's cervical spine indicate that the surviving skull fragment was sawn when Kelly's head was attached to his neck. The saw marks on the skull fragment are not indicative of the usual autopsy practice in removing a brain – the horizontal saw mark is too low.
We will never really know the truth of this, unless the remainder of the skull is found or the full memorandum by Castieau to the police is unearthed.
4 Kelly's head was removed for the purpose of making the death mask.
There is no evidence that a prisoner's head would be removed in order to make a death mask. The historical information on the making of death masks describes a process by which the mould for the mask is made on a complete body (see Fig. 25.2). The curator of the Australian Institute of Anatomy has confirmed this view in correspondence on the making of Kelly's death mask. There is also an example of a complete skeleton in the mortuary believed to belong to a prisoner who had a death mask made: (Phelan – coffin marked with his name).
5 Professor Richard Berry said that Deeming, the murderer, had a brain the development of which had been arrested when he was aged 13 years and that Ned Kelly was only slightly more intelligent, having a 14-year-old brain development. His examinations of criminals in the Melbourne Gaol showed that the average cubic capacity of their brains was 100 cubic cm less than normal. He must have had access to the brains of those criminals in making that study.

Fig. 25.1: The death mask of Ernest Knox from the Old Melbourne Gaol. Ronn Taylor.

Fig. 25.2: Two men making a death mask, New York, c. 1908. Wikipedia.

Professor Berry made his cubic capacity calculations from both the heads of living criminals and the death masks of criminals. He used an algorithm to calculate the capacity of the brain from the measurements of the heads. He has published how he did this.

6 A senior member of the Catholic clergy had buried Kelly's skull in consecrated ground.
This could be possible if someone souvenired the skull in 1929 and gave it to a member of the Catholic Church – we may never know.

7 Kelly's skull is in the basement of the Australian Institute of Anatomy.
This skull was the 'Baxter skull' that was later stolen from the Old Melbourne Gaol Museum. The VIFM has shown through DNA analysis that this skull is not that of Kelly, but of another prisoner buried at the Melbourne Gaol.

8 The skull was used as a paperweight in the gaol or by a government official.
When Harry Lee took the skull believed to be Kelly's from the Melbourne Gaol site in April 1929 (see Chapter 2), he kept it at his house for a couple of days before handing it over to the authorities. We know that it ended up at the Australian Institute of Anatomy (AIA) in the early 1930s, but where was it between April 1929 and then is unaccounted for. It may well have sat on some police officer's or government official's desk, as the circumstances under which it ended up at the AIA have not been documented. This skull, however, was the 'Baxter skull', not Ned Kelly's.

References

Dimond J (2013) Ned Kelly's skull. *Overland* **60**, 211.

Chapter 26
Solving the mystery of the skull

David Ranson and Deb Withers

In a quiet and quaint cemetery in Merseyside, not unlike those which feature in the British television crime series *Midsomer Murders*, there lie the remains of a man who could provide the last piece of the jigsaw in the identification of the skull long thought to belong to Ned Kelly.

His name was Thomas Bailey and he was the brother of convicted multiple murderer Frederick Bailey Deeming (see p. 117).

Thomas is buried under the maiden name the family chose to avoid Frederick Deeming's notoriety. Deeming murdered his wife and four children in Rainhill, England, left them buried under the fireplace, and made his way to Australia, where he murdered his second bride and buried her under the fireplace in a house that still exists in the Melbourne suburb of Windsor.

Circumstantial evidence that the 'Baxter' skull belonged to Deeming was quite strong, as has been shown in other chapters in this book (see Chapters 6 and 9). Deeming's death mask was a close fit to the skull when craniofacial superimposition was applied, and despite the 12-year gap from Kelly's execution in 1880 and Deeming's in 1892, records indicated that Deeming's grave in the Old Melbourne Gaol cemetery could have been alongside Ned Kelly's.

But the only way to know for sure was to get a DNA sample from a relative of Deeming who shared his DNA and then compare it to the DNA which the VIFM team had already obtained from the skull.

Through the research of a British genealogist, Thomas' great, great grand-daughter Kathleen Young, of Churchtown, gave the VIFM her permission to exhume her ancestor for sampling. Following due process the Victorian Attorney General wrote to his British counterpart in the Home Office, who agreed to help, and the cemetery trust also granted permission.

The final approval was then obtained from the Church of England Archdiocese. This was almost unprecedented as the Church has strict rules about a person's final resting place and rarely grants permission for exhumation.

The exhumation finally took place on Tuesday May 6th, 2014. At midnight on Tuesday May 6th, A/Prof David Ranson, forensic pathologist and deputy director of the VIFM, and Dr Noel Woodford, forensic pathologist and head of forensic pathology at the VIFM, met cemetery registrar Lisa Parkes at Bebington Cemetery, about a 30-minute drive from Liverpool.

There was a chill in the air befitting the scene and soon it was to rain. Lisa and her team had begun preparing the excavation of Thomas' grave earlier in the day. They had erected a tent for privacy and protection from the weather and installed portable lights to assist David and Noel in a task that was not rare but was certainly rather unusual.

The exhumation process had to be conducted with due regard for decency and privacy and at the same time allow for uncontaminated DNA samples to be obtained. The archaeology of the grave and those buried within it had to be carefully researched as it was common for more than one body to be buried in a plot, and for families to be buried close to each other at different layers within the grave. If Thomas Bailey's children were buried there as well, and since they wouldn't have inherited the mtDNA of their father, their remains would have to be identified and excluded.

Cemetery records indicated that the grave contained the remains of 63-year-old Thomas, his wife Susan, their four-year-old son Reginald and 26-year-old daughter Suzanne. As Susan died last, in 1935, she had to be exhumed before Thomas, who died in 1911. It was imperative that the sample only be taken from Thomas' remains.

The body of Susan and her partly disintegrated coffin were carefully removed from the grave exposing the remains of Thomas's coffin lid, beneath which lay his bones. It was clear that these remains were of a mature adult male and each bone was carefully removed from the grave together with some teeth from adjacent to his skull. The teeth and the femurs were laid out on a trestle table for collection of the biopsies for DNA extraction. A range of biopsies were collected to ensure that the appropriate analysis could be undertaken. Much depended on the condition of the remains. If they had been allowed to dry out in sandy soil with the disintegration of the casket then DNA preservation would be optimum; however, if they were very wet or had water drainage flow through the bones, obtaining DNA from the samples could have been trickier.

Fortunately Thomas' grave was on high, dry land, near leafy trees and a footpath.

Proper packaging and transport would then prevent further degradation or subsequent contamination of the remains.

The bone biopsy samples were collected using a small bone saw with separate blades, first washed with bleach and alcohol, before each cut. Following the collection of the samples, both Susan and Thomas's bones were separately re-coffined and reinterred with appropriate ceremony in conjunction with the Church and cemetery requirements.

After packing the samples in sterile containers according to protocol, David brought the samples back to the VIFM, and they were given to Dr Dadna Hartman, head of molecular biology for analysis.

In the DNA laboratory, the most suitable samples were cleaned following established protocols to remove external contaminants and DNA extracted so that an mtDNA profile could be established. Once a profile was successfully obtained from the remains, this became a reference or comparison profile. If the skull was indeed that of Frederick Deeming, then its mtDNA profile would be identical to the profile obtained from these remains.

Dadna and her team obtained a profile but to ensure their work was in no doubt, she also asked her colleagues at the Australian Centre for Ancient DNA laboratory in Adelaide conduct the same tests to ensure they obtained the same results. In order to prove there had been no DNA contamination at the time of the graveside biopsies, both David and Noel provided exclusion mtDNA samples and fortunately their mtDNA was not detected in the tooth or bone samples.

Everyone involved in the Kelly project now had to wait patiently for the final results. It would take three weeks but the wait was worth it for those who had spent so long working to identify Ned Kelly's remains.

Unfortunately, the mtDNA from the 'Baxter skull' – which had been exhumed in 1929, spent time on a bedside table, used as a paperweight and eventually placed on display in a glass case next to Ned Kelly's death mask in the Old Melbourne Gaol Museum – did not match the mtDNA from the bones of the brother of Frederick Deeming. So the mystery continues. The 'Baxter skull' is not that of Ned Kelly. It is not that of Frederick Deeming. But it is certainly that of an executed prisoner who had been buried at the Old Melbourne Gaol. Not all the executed prisoners had death masks that were able to be found, not all the prisoners' remains had skulls, not all the remains were able to provide satisfactory DNA for analysis. Will the skull ever be identified? The answer to that is unclear. Forensic science and its associated technology are improving all the time so ... watch this space!

Afterword

The Hon. Robert Clark
Attorney-General of Victoria (2010–)

In 1880, Victoria had a population of 450 558, Graham Berry was having his third (brief) stint as Premier, William Stawell, who had been Victoria's first Attorney-General in 1851, was well into his third decade as Chief Justice. Grand Flaneur won the Melbourne Cup and Geelong its third premiership in a row. On 11 November that year, Ned Kelly was executed.

Among all the historical figures in our state's history, Kelly is unique in his ability to generate debate and polarise opinion. To many, he was and remains a romantic figure who defied unjust authority. To others, he was a notorious thief and cold-blooded killer. In that respect, Ned Kelly can rightly be regarded as the most controversial individual that Victoria has produced.

The initiative by VIFM to take on the task of identifying the remains of Ned Kelly from among the many prisoners exhumed from Pentridge Prison has been strongly supported by successive Victorian governments.

The identification of his remains has focused attention on several questions associated with Kelly's life, his death and the controversy he has always generated.

Upon his death, Kelly was buried in the Old Melbourne Gaol, along with other executed prisoners. In 1929 the burial site was excavated during redevelopment works at the Melbourne Gaol and Kelly's coffin was inadvertently broken open. When news of the incident spread, a large unruly mob gathered at the site. Several individuals scrambled into the grave in order to souvenir various fragments of the skeleton. A report in the *Age* described the events as a 'scandalous exhibition of ghoulish depravity' and condemned interest in the remains as 'the morbid desire to possess ghoulish relics of the notorious bushranger' (*Age*, 16 April 1929).

Another newspaper, the *Truth*, took a different view and very prophetically wrote that:

> Fate seems determined that Australia shall not be allowed to forget him, for even after 50 years the light of publicity which has never left his name flares up afresh through the dismemberment of his skeleton (*Truth*, 20 April 1929).

The ultimate identification of the remains was the culmination of 20 months of ground-breaking work by VIFM and its partners. This process has developed VIFM's expertise in ways that will benefit Victoria for years to come. It also involved collaboration with world-leading forensic experts from other jurisdictions. In this regard, I wish to particularly acknowledge the involvement of the DNA Laboratory of the Argentine Forensic Anthropology Team, with whom the Institute has worked in East Timor as well as collaborations elsewhere.

The identification of the remains of Ned Kelly is not only an event of great professional significance for VIFM, it is a significant new chapter in Victoria's most famous historical drama. I heartily congratulate all of the VIFM staff and other members of the team who contributed to this remarkable achievement.

Appendix 1: DNA processes

Dadna Hartman and Carlos Vullo

Polymerase chain reaction (PCR)

A process of amplification (known as polymerase chain reaction or PCR) is used to make 100s of copies of the HVI and HVII (hypervariable 1 and hypervariable 2) regions of the mtDNA – similar to using a photocopying machine to make multiple copies of one page.

These amplified fragments of the HVI and HVII regions are then sequenced to determine the DNA composition/order for each fragment (see Figs App1.1, App1.2).

DNA is made up of four building blocks known as A, G, T and C. The order of these give rise to a unique sequence of DNA – similar to the letters of the alphabet coming together in a particular sequence to make words and sentences. By determining the unique order, or sequence, of the DNA, we can determine the mtDNA profile. The profile is reported as the noted differences from a standard reference, which is the Cambridge mitochondrial reference sequence.

mtDNA profile comparisons

What is an mtDNA profile? In simple terms, it is the unique sequence of DNA that can be attributed to the sample analysed. It is presented as the noted differences between the sequence of the sample and that reported as the Cambridge Reference Sequence (CRS) – which is the first reported DNA sequence of a mitochondrial genome published in the scientific community. Using computer software programs that are capable of aligning your query mtDNA sequence with that of the CRS, you can determine where the differences (if any) occur. The difference are then reported based on their location in the mtDNA genome and their composition, in a table format (see Fig. App1.3).

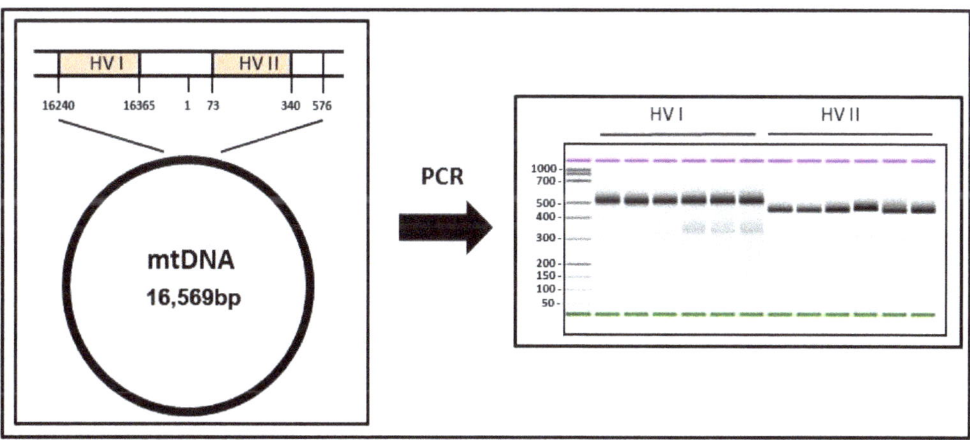

Fig. App1.1

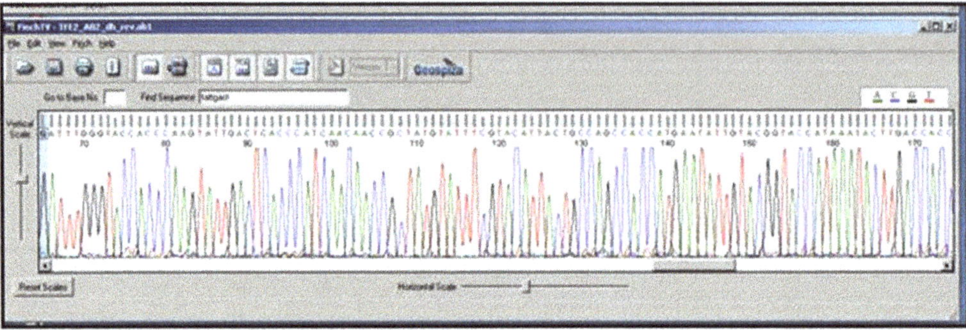

Fig. App1.2

Using the DNA comparison software, all the sequences to be compared are first imported into the program (panel A). In the example given, sample 1 and positive control have been analysed in duplicate and are compared to the CRS. The software then highlights the regions that differ between the samples and the CRS (panel B, arrows indicate differences to the CRS). The positions at which differences are found (e.g. 16224 and 16292) are noted and reported in the table as shown. The mtDNA profile for sample 1 (in the example given) would be 16224C, 16311C and 16320T.

mtDNA haplotypes, haplogroups and human genetics

What can an mtDNA profile tell us? Although 16224C, 16311C and 16320T may seem like someone's home phone number, as an mtDNA profile (or haplotype, as it is also known) it

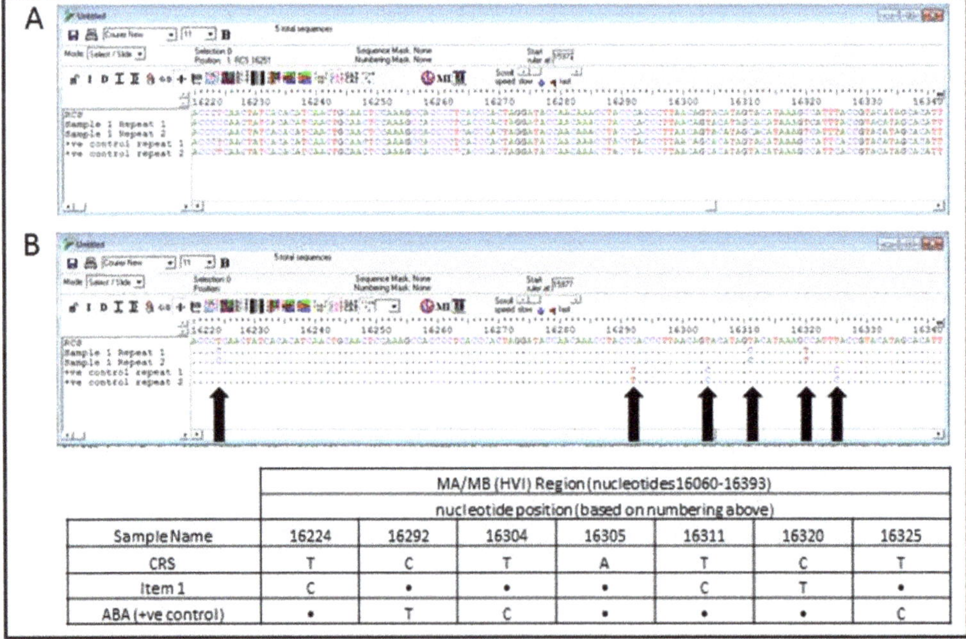

	MA/MB (HVI) Region (nucleotides16060-16393)						
	nucleotide position (based on numbering above)						
Sample Name	16224	16292	16304	16305	16311	16320	16325
CRS	T	C	T	A	T	C	T
Item 1	C	•	•	•	C	T	•
ABA (+ve control)	•	T	C	•	•	•	C

Fig. App1.3

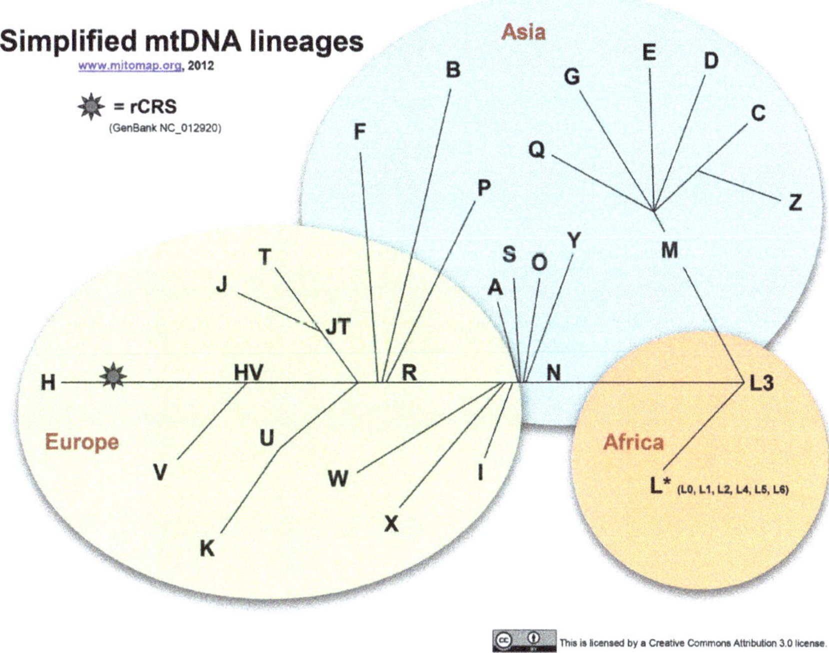

Fig. App1.4

can tell us a great deal. Haplotypes are further grouped into haplogroups. These classifications can be used to ascertain the origins or heritage of individuals who have that particular haplotype.

Scientists have been studying the mtDNA haplotypes of humans for many years. Given its unique inheritance through the maternal lines, evolutionary studies can trace the matrilineal inheritance of modern humans back to its origins in Africa to the most recent common ancestor – commonly called Mitochondrial Eve, who lived approximately 140 000 to 200 000 years ago.

Similar haplotypes are grouped together into a haplogroup, the way one might group granny smiths, jazz or red delicious apples into the one group. Haplogroups are used to represent the major branch points on the mitochondrial phylogenetic tree and have been assigned the letter A–Z, each with many sub-branches. Some haplotypes within a haplogroup are more common than others, and they segregate to different parts of the world based on past migration patterns (see Fig. App1.4).

Going back to the example haplotype, 16224C, 16311C and 16320T, we know this to belong to the haplogroup K1c2 (descendants from K) most commonly found in Europe.

Table App1.1: mtDNA profiling results for the Pentridge Prison bone samples

Pentridge Prison remains	HG	Haplotype	mtDNA results
Pen1	n/d	16261T, 200G, 263G, 315.1C	HV1/2
Pen2	D4j*	16223T, 16291T, 16362C, 73G, 263G, 309.1C, 315.1C	HV1/2
Pen3	T2b	16126C, 16294T, 16296T, 16304C, 73G, 263G, 309.1C, 309.2C, 315.1C	HV1/2
Pen4	H	210G, 315.1C	HV1/2
Pen5	U5a	16256T, 16270T, 16291T	HVI only
Pen6			not reportable
Pen7	T2b	16126C, 16266T, 16294T, 16304C, 73G, 198T, 263G, 309.1C, 315.1C	HV1/2
Pen8	H1b	16183C, 16189C, 16356C, 16360T, 263G, 315.1C	HV1/2
Pen9	U8a1	16146G, 16342C, 73G, 263G, 282C, 309.1C, 315.1C	HV1/2
Pen10	H	16278T, 153G, 263G, 315.1C	HV1/2
Pen11	T2	16126C, 16294T, 16296T, 73G, 263G, 315.1C	HV1/2
Pen12			not reportable
Pen13	H	16256T, 263G, 315.1C	HV1/2
Pen14	**J1c**	**16069T, 16126C, 73G, 228A, 263G, 295T, 315.1C**	**HV1/2**
Pen15	H	93G, 263G, 315.1C	HV1/2
Pen16	H	16148T, 263G, 309.1C, 309.2C, 315.1C	HV1/2
Pen17	n/d	73G, 125C, 127C, 150T, 152C, 263G, 309.1C, 315.1C	HV1/2
Pen18	n/d	16183G, 16223T, 16259T, 16357C	HV1 only
Pen19	T1a	16126C, 16163G, 16186T, 16189C, 16294T, 73G, 152C, 195C, 263G, 315.1C	HV1/2
Pen20	HV0*	16162G, 16295T, 16298C, 263G, 309.1C, 315.1C	HV1/2
Pen21	P4b	16274A, 16291T, 16337T, 16352C, 16362C, 73G, 263G, 315.1C	HV1/2
Pen22			not reportable
Pen23	H5	16304C, 228A, 263G, 315.1C	HV1/2
Pen24	U5b1	16174T, 16189C, 16270T, 16311C, 73G, 150T, 263G, 315.1C	HV1/2
Pen25	H1a	16051G, 16162G, 73G, 263G, 315.1C	HV1/2
Pen26	M4c	16072T, 16184T, 73G, 152C, 195C, 263G, 315.1C	HV1/2
Pen27	HV0*	16162G, 16295T, 16298C, 263G, 309.1C, 315.1C	HV1/2
Pen28			not reportable
Pen29	K	16119G, 16224C, 16311C, 73G, 263G, 309.1C, 315.1C	HV1/2
Pen30			not reportable
Pen31	H5	16304C, 146C, 263G, 309.1C, 315.1C	HV1/2
Pen32			not reportable
Pen33	HV0*	16298C, 16311C, 263G, 309.1C, 309.2C, 315.1C	HV1/2
Pen34	H	263G, 315.1C	HV1/2

Pen35	H	263G, 315.1C	HV1/2
Pen36			not reportable
Mitochondrial DNA analysis			**% success rate (# cases)**
Complete profile			75 (27)
Partial profile			5.5 (2)
Not reportable			19.5 (7)

HG = haplogroup.
n/d = haplogroup not determined.
Highlighted line corresponding to Pen14 (case 3081/10) remains that matched to MRNK.
Source: Blau *et al.* (2014).

Table App1.2: Mitochondrial haplogroup distribution (%); Western Europe [2] compared to the Pentridge Prison cases

Haplogroup	Western Europe (%)*	Pentridge (%)
D	1	3.7
H	41	40.8
I	2	-
J	9	3.7
K	5	3.7
L	1	-
M	1	3.7
N	1	-
T	8	14.8
U	18	7.4
V/HV0	7	11.1
W	2	-
X	2	-
Other	3	11.1

Source: Blau *et al.* (2014).

Appendix 2: Metal crystallography

Gordon Thorogood

The micrograph in Fig. App2.1, from the back plate, is a typical example of grain deformation due to the metal being worked below its recrystallization temperature (see also Figs App2.2, App2.3). An acetate replica taken edge-on from the breast plate (see photo), however, showed a banded structure. This is due to the original rolling of the metal and indicates that the bulk of the metal has not been heated above the recrystallization temperature.

Another replica taken from the breast plate showed evidence of pearlite being adjacent to oxide corrosion (see Figs App2.4, App2.5). This showed the metal had been at a temperature

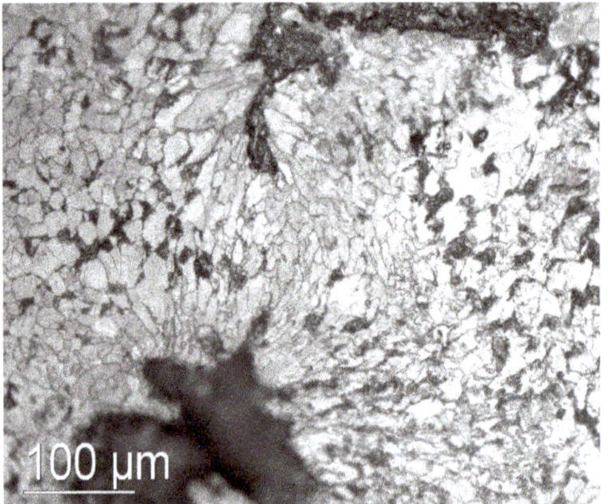

Fig. App2.1: Note deformed grains around holes indicating working of the metal below the recrystallisation temperature. ANSTO.

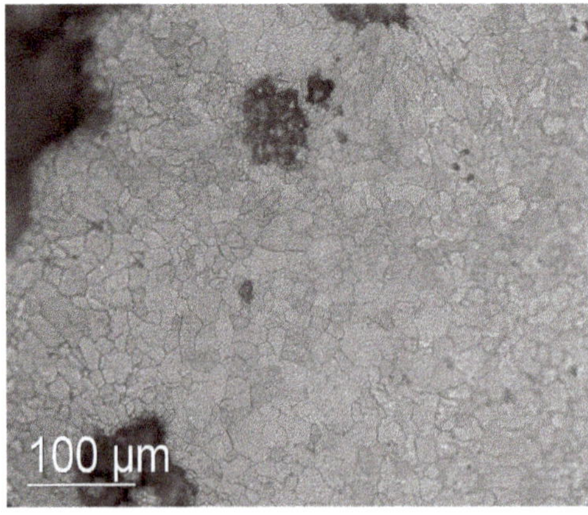

Fig. App2.2: Area adjacent to Fig. App2.1. Note 100% ferrite structure indicating decarburisation during hot working. ANSTO.

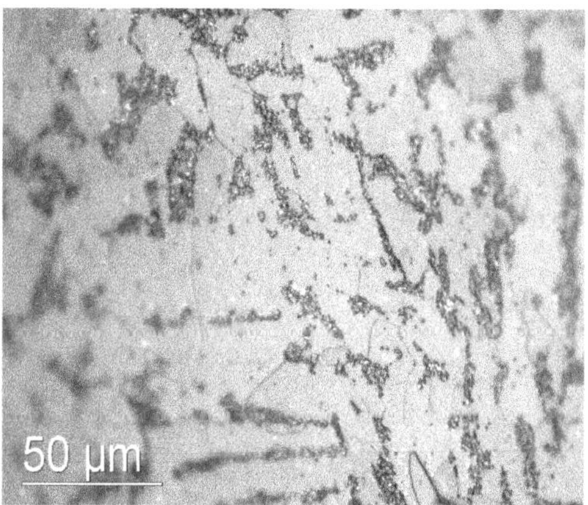

Fig. App2.3: Spheriodised carbides are evident in thin section of metal. ANSTO.

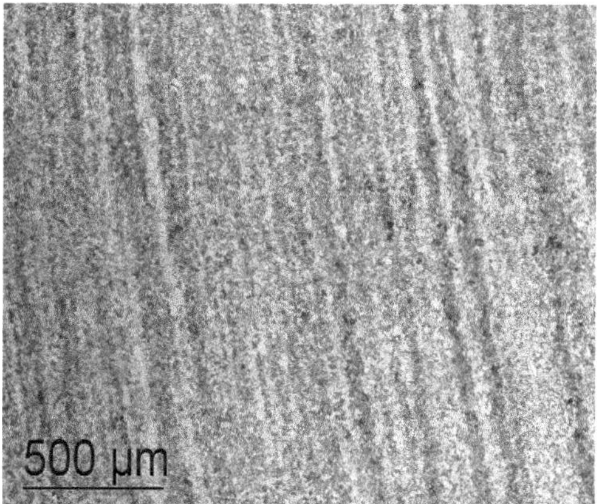

Fig. App2.4: Replica taken from lower corner edge of breast plate. Banded structure is due to rolling of original metal. ANSTO.

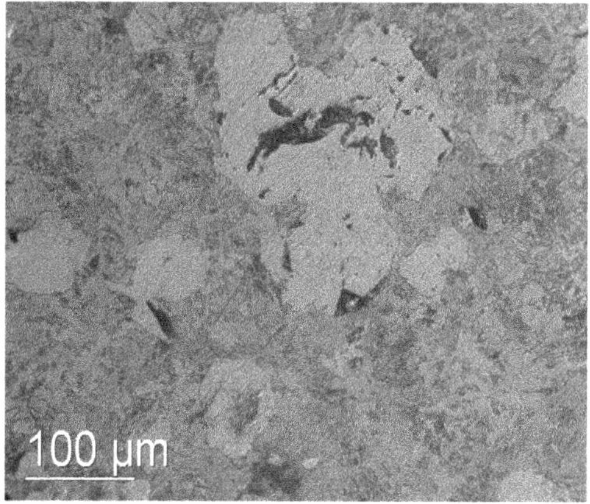

Fig. App2.5: Note the adherent nature of oxide to pearlite grains. ANSTO.

sufficiently high to oxidise the surface but not high enough to recrystallize the surface, as pearlite grains are adjacent and adhering to the oxide. This behaviour is consistent with this part of the armour being heated in the temperature range of 500–700°C.

Replicas taken from the helmet (see Figs App2.6, App2.7) showed an uneven grain size but the pearlite had taken on the spheriodised structure. This indicated that the helmet had been heated close to the recrystallization temperature but not above it. And a replica taken from an adjacent area (see Fig. App2.8) showed sulphide stringers. These were due to the original rolling of the metal, though, and again provide evidence that the steel had not been heated to a temperature high enough to modify the original metallurgical structure.

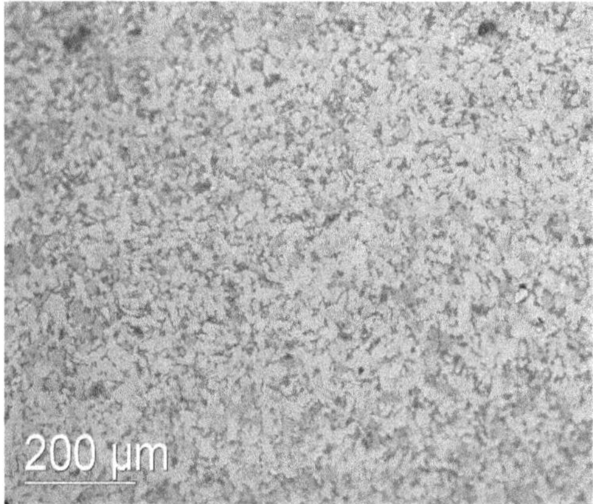

Fig. App2.6: Top edge of helmet. Note fine and slightly uneven grain size and shape. Pearlite is evenly distributed throughout the structure. ANSTO.

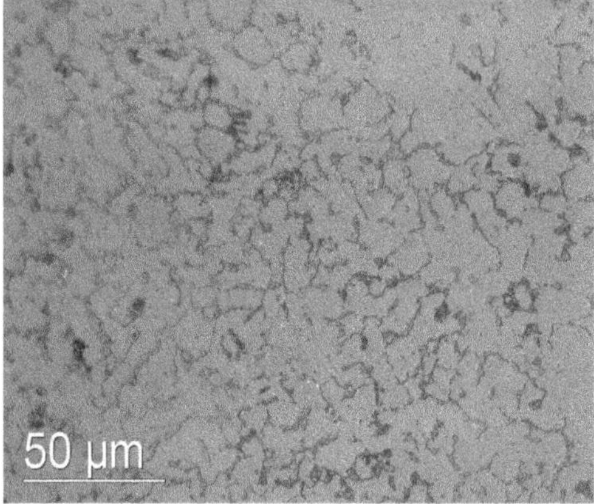

Fig. App2.7: Replica from helmet with spheriodised structure of pearlite. ANSTO.

Table App2.1: Analysis of particles labelled in Fig. App2.10

Particle	Composition
1	Mn + S
2	Fe + trace Mn
3	Mn + S
4	Fe + trace Mn
5	Fe + trace Mn
6	Fe + trace Mn
7	Fe + trace Mn
8	Fe + trace Mn

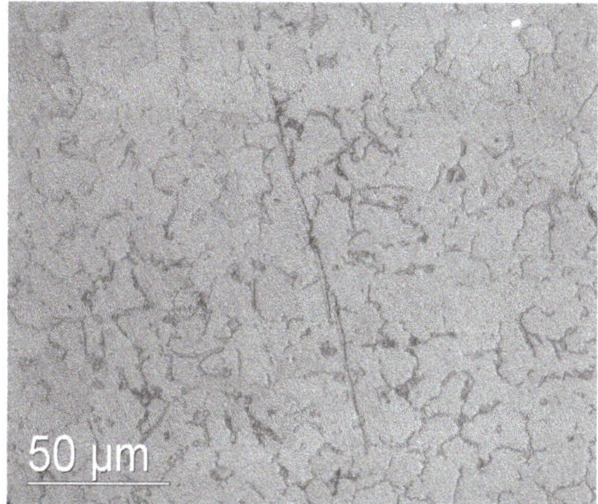

Fig. App2.8: Centre of image shows sulphide stringers, which are an original rolling artefact. ANSTO.

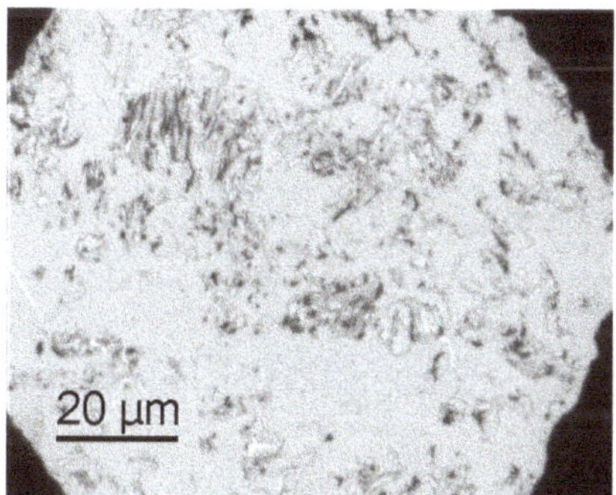

Fig. App2.9: TEM extraction replica – low magnification. Typical region showing a small pearlite colony (top left) consisting of parallel laths of Fe_3C (cementite). A large number of failed extractions, particularly of the larger carbides, is evident. Large areas devoid of carbides are ferrite grains. ANSTO.

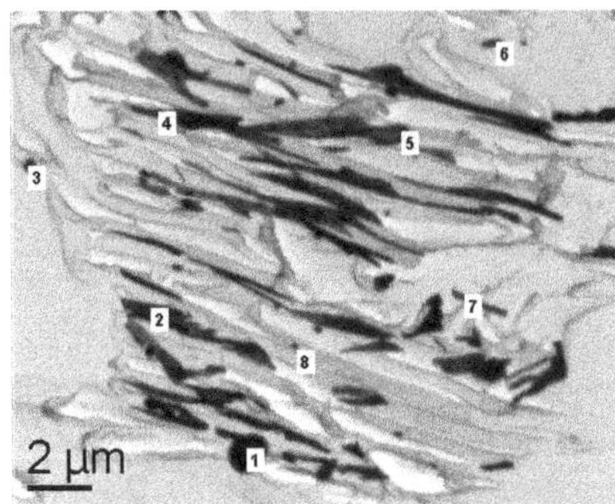

Fig. App2.10: Detail of pearlite colony. Several larger laths of Fe_3C have failed to extract. The extracted laths have a high aspect ratio and sharp ends, suggesting that spheroidisation is minimal. A large rounded MnS inclusion is present at the bottom of the colony (1)s. Numbered positions relate to EDS point analyses (see text). ANSTO.

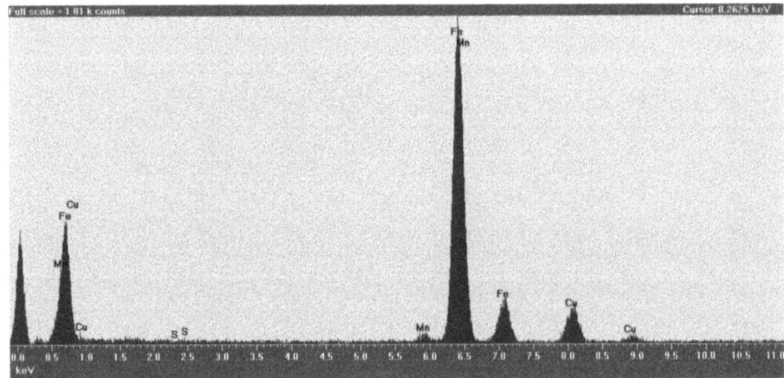

Fig. App2.11: EDS spectrum from particle 2 in Fig. App2.10. The analysis is characteristic of Fe_3C, with small amounts of manganese dissolved in it. The copper peak is an artefact caused by the copper grid on which the replica is mounted. ANSTO.

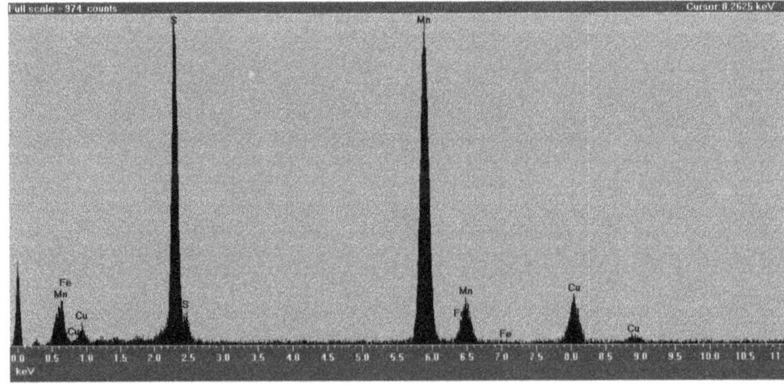

Fig. App2.12: EDS spectrum from particle 1 in Fig. App2.10. The analysis is characteristic of MnS. Manganese is added to the steel during manufacture to de-oxidise it. This element also has the advantage of locking up undesirable sulphur into globular inclusions. ANSTO.

Index

www.ingramcontent.com/pod-product-compliance
Lightning Source LLC
LaVergne TN
LVHW060629110826
845147LV00014B/879
* 9 7 8 1 4 8 6 3 0 1 7 6 8 *